The Ordovician Deicke and Millbrig K-Bentonite Beds of the Cincinnati Arch and the Southern Valley and Ridge Province

John T. Haynes*
Department of Geography and Earth System Science
George Mason University
Fairfax, Virginia 22030-4444

290

1994

*Present address: Department of Mineral Sciences, The Smithsonian Institution, Washington, D.C. 20560.

Published by The Geological Society of America, Inc.
3300 Penrose Place, P.O. Box 9140, Boulder, Colorado 80301

Printed in U.S.A.

GSA Books Science Editor Richard A. Hoppin

Library of Congress Cataloging-in-Publication Data

Haynes, John T.
The Ordovician Deicke and Millbrig K-bentonite Beds of the Cincinnati Arch and the southern valley and ridge province / John T. Haynes.
p. cm. — (Special paper ; 290)
Includes bibliographical references.
ISBN 0-8137-2290-X
1. Bentonite—Southern States. 2. Geology, Stratigraphic--Ordovician. 3. Geology—Southern States. I. Title. II. Series: Special papers (Geological Society of America) ; 290.
QE471.15.B4H39 1994
551.7'31'0975—dc20 94-3151
CIP

Front cover: *Top left:* The Deicke K-bentonite Bed in the Carters Limestone at South Carthage, Tennessee. Photograph taken by Warren D. Huff. *Top right:* The Deicke K-bentonite Bed in the Eggleston Formation at Hagan, Virginia. Bedding top to left. Photograph taken by Warren D. Huff. *Bottom left:* The Millbrig K-bentonite Bed at the Tyrone-Lexington contact, Shakertown, Kentucky. Photograph taken by Warren D. Huff. *Bottom right:* The Millbrig K-bentonite Bed in the Bays Formation at Crockett Cove, near Wytheville, Virginia. Photograph taken by John T. Haynes. **Back cover:** *Left:* Feldspars and opaque ferro-titanian minerals in the Deicke K-bentonite Bed. Core from DeKalb County, Tennessee. Thin section, cross-polarized light, x3. *Right:* Quartz, biotite, and feldspars in the Millbrig K-bentonite Bed, from Big Ridge, Alabama. Thin section, cross-polarized light, x3.

This book is dedicated to all the geologists who have worked on Ordovician K-bentonites in the Appalachians.

10 9 8 7 6 5 4 3 2 1

Contents

Geological Society of America
Special Paper 290
1994

The Ordovician Deicke and Millbrig K-Bentonite Beds of the Cincinnati Arch and the Southern Valley and Ridge Province

ABSTRACT

In the southeastern United States, identification of the Rocklandian Deicke and Millbrig K-bentonite Beds is based on differences in phenocryst mineralogy in the tuffaceous zones of each bed, and the two beds can be reliably and consistently distinguished on this basis. The common phenocrysts in the Deicke are labradorite and various Fe-Ti minerals, and in the Millbrig they are andesine, quartz, and biotite. Phenocrysts present in trace amounts are biotite and quartz in the Deicke, and apatite and zircon in both beds. The Deicke is altered dacitic or latitic ash, whereas the Millbrig is altered rhyodacitic ash. Both beds are interpreted as airfall deposits produced by huge volcanic eruptions, each of which was much larger and produced far more ash than the 1815 eruption of Tambora and even the great Toba eruption of 75 Ka.

The principal authigenic minerals are feldspars (albite and K-feldspar), the clay minerals mixed-layer illite/smectite (I/S) and kaolinite, and ferroan and ferrotitanian minerals (pyrite, hematite, and the TiO_2 polymorphs). The current lateral distribution of these authigenic minerals is the result of variations in regional geochemical conditions during burial diagenesis. The distribution of authigenic feldspars, together with regional changes in the percentage of illite in the I/S, indicates that maximum burial temperatures were higher in the Valley and Ridge province than along the Cincinnati Arch, and that the highest temperatures were in the Virginia Valley and Ridge. Regional variations in the distribution of the ferroan and ferrotitanian authigenic minerals in the K-bentonites and adjacent strata reflect variations in pore water redox conditions during diagenesis. Throughout the Cincinnati Arch and the westernmost Valley and Ridge, sediment pore waters were reducing. In the eastern Valley and Ridge of Alabama and Georgia the pore waters were oxidizing, and in the central and eastern Valley and Ridge from south of Roanoke to Knoxville the pore waters were initially reducing and then oxidizing.

The stratigraphy of the Deicke and Millbrig is now very well known from their type area in the Upper Mississippi Valley to the easternmost exposures of Rocklandian strata in the southern Valley and Ridge from Alabama to Virginia. Many previous correlations are verified, and correlation of the two beds is extended along and across strike in the Valley and Ridge between Roanoke, Virginia, and Birmingham, Alabama. The Deicke and Millbrig are the thickest and most wide-

Haynes, J. T., 1994, The Ordovician Deicke and Millbrig K-Bentonite Beds of the Cincinnati Arch and the Southern Valley and Ridge Province: Boulder, Colorado, Geological Society of America Special Paper 290.

spread of the several Rocklandian K-bentonites in this region. Of the two, the Deicke is more laterally persistent along the Cincinnati Arch, but in the southern Valley and Ridge the Millbrig is the more widespread and persistent of the two beds. The relationship of the K-bentonites to several regional and local unconformities is also now better understood. With this documentation of the great lateral extent of these two superb marker beds, it will be possible to link the vertical and lateral distribution of stratigraphically important faunal elements, especially conodonts and graptolites, with the stratigraphic position of the Deicke and Millbrig K-bentonite Beds.

INTRODUCTION AND PURPOSE

This publication reports on an investigation of the petrology, diagenetic history, and stratigraphy of the Ordovician Deicke (pronounced Di´kē) and Millbrig K-bentonite Beds based on study of K-bentonites and adjacent strata at more than 60 outcrops and cores along the Cincinnati Arch and in the southern Appalachians of the southeastern United States (Fig. 1). The Deicke and Millbrig, formally named by Willman and Kolata (1978), are two of the many beds of altered volcanic ash that occur in Rocklandian (Middle to Upper Mohawkian/Champlainian [Lower "Trentonian" of older usage], = Lower to Middle Caradoc) strata of eastern North America. Biotite, quartz, feldspars, and other nonclay minerals are present as primary phenocrysts and secondary phenoclasts in the Deicke and Millbrig in the study area—and, in fact, are so abundant and coarse grained in some zones that the material in those zones is tuffaceous rather than bentonitic—but the principal constituent of these and many other such beds in Lower Paleozoic strata is mixed-layer illite/smectite (I/S) clay that generally contains 70 to 90% illite. With this much illite in the crystal lattice of the I/S, the clay of these K-bentonites is only moderately expandable. Thus these particular beds of altered volcanic ash are correctly identified as potassium bentonites or K-bentonites (Weaver, 1953) rather than as true bentonites, metabentonites, or ash beds (Huff, 1983a; Forsman, 1984; Haynes and Huff, 1990).

Many localities where two or three Ordovician K-bentonites that are markedly thicker and coarser grained than other K-bentonites in the exposure occur in the southeastern U.S., and some are quite well known, for example, Hagan, Virginia, and Big Ridge, Alabama. Several informal stratigraphic nomenclatural systems that attempted to organize K-bentonite stratigraphy systematically around these thicker and more laterally persistent beds have been introduced into the literature. They include the alpha-numeric V, B, T, R, N, and S systems of Rosenkrans (1936), Fox and Grant (1944), Wilson (1949), Miller and Fuller (1954), Kay (1944, 1956), and Perry (1964), respectively. In those systems each K-bentonite at the "type" locality described or cited in the publication was given a sequential alpha-numeric name such as B-1, B-2, B-3, etc., depending on the location of the bed in the measured section. Some of the alpha-numeric names have been widely used for the thicker and more laterally persistent beds, for example, T-3 and T-4 in Tennessee, Alabama, and Georgia (Milici, 1969; Chowns and McKinney, 1980), and V-3, V-4, and V-7 in southwestern Virginia (Huffman, 1945; Hergenroder, 1966). From these and other studies, the Mohawkian K-bentonite stratigraphic framework is now understood locally in a general sense for Virginia (Rosenkrans, 1936; Huffman, 1945; Miller and Brosgé, 1954; Miller and Fuller, 1954; Kay, 1956; Hergenroder, 1966; Haynes, 1992), Tennessee (Fox and Grant, 1944; Wilson, 1949, 1991; Milici, 1969), Kentucky (Young, 1940; McFarlan, 1943), Georgia (Allen and Lester, 1957; Milici and Smith, 1969; Carter and Chowns, 1983a, 1986), and Alabama (Drahovzal and Neathery, 1971; Chowns and McKinney, 1980).

In their type areas in the Upper Mississippi Valley, the Deicke and Millbrig are recognized as formal stratigraphic units, in this case as "Beds" (Willman and Kolata, 1978). Deicke K-bentonite Bed and Millbrig K-bentonite Bed are terms that satisfy the requirements of the North American Stratigraphic Code (North American Commission on Stratigraphic Nomenclature, 1983), and the use herein of the formal names Deicke K-bentonite Bed and Millbrig K-bentonite Bed is thus appropriate, although the various alpha-numeric names (such as T-3 and T-4) can continue to be useful informal designations in much the same way, for example, that "Number 9 Coal" is a useful designation in certain Pennsylvanian strata. In parts of the present

Figure 1. Outcrop of Ordovician strata (stippled and outlined) along the Cincinnati Arch and in the southern Valley and Ridge from Roanoke, Virginia, to Birmingham, Alabama, showing the location of the principal surface exposures and cores studied. The section numbers refer to the locality register of Appendix 1. The two broad areas of outcrop in the vicinity of Lexington, Kentucky, and Nashville, Tennessee, are associated with the Jessamine and Nashville Domes, respectively; shown is the approximate areal extent of the High Bridge Group and Lexington Limestone (Kentucky) and the Stones River and Nashville Groups (Tennessee). The outcrop areas in the southern Valley and Ridge show the approximate areal extent of the post-Knox, pre-Oswego/Juniata/Sequatchie stratigraphic interval. Areal geology adapted from state geologic maps of Kentucky, Virginia, West Virginia, Tennessee, Georgia, and Alabama.

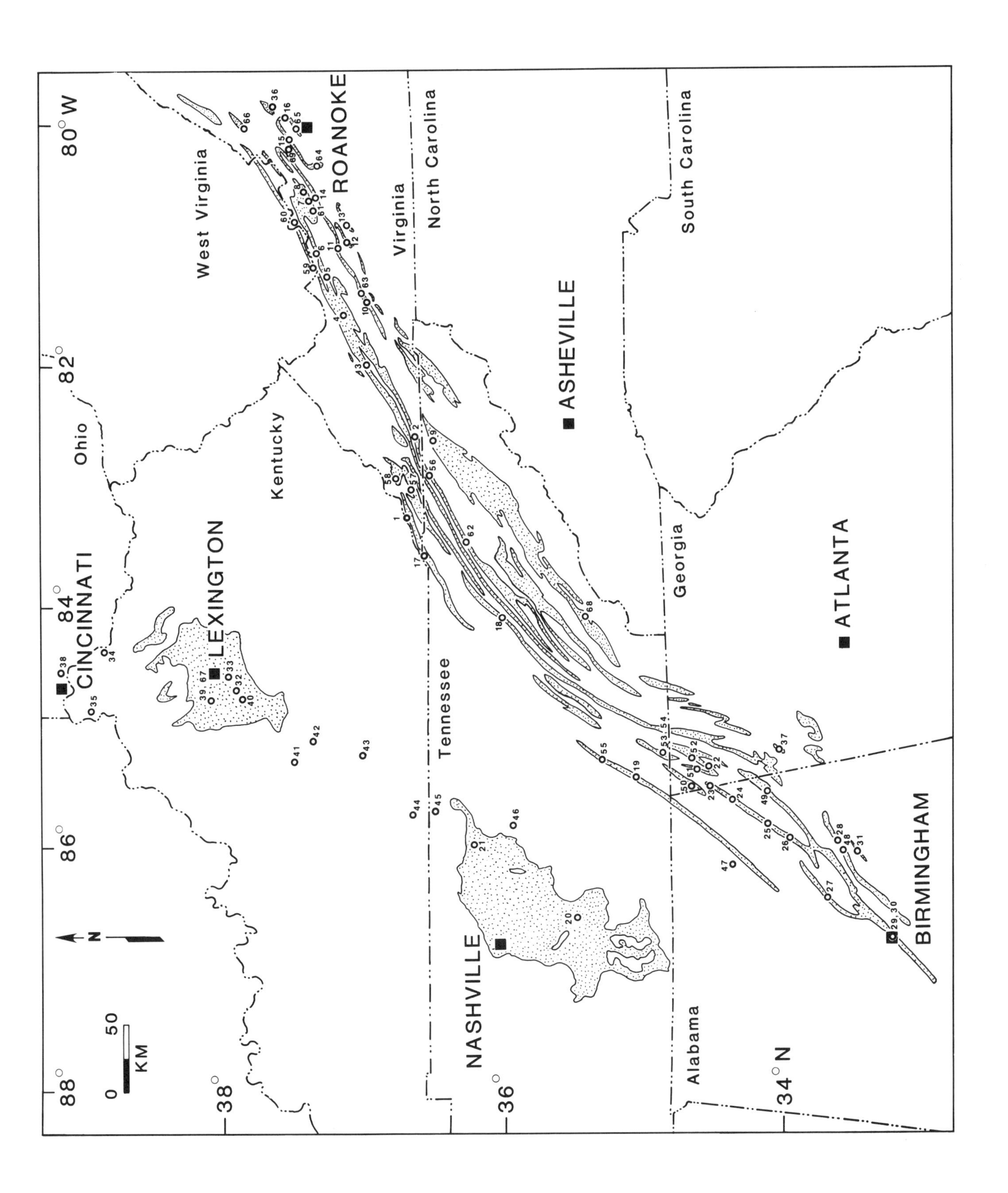

West Virginia
Virginia
North Carolina
South Carolina
Ohio
Kentucky
Tennessee
Georgia
Alabama
CINCINNATI
LEXINGTON
ROANOKE
ASHEVILLE
ATLANTA
NASHVILLE
BIRMINGHAM
80° W
82°
84°
86°
88°
38°
36°
34° N
N
0
50
KM

publication the alpha-numeric names are used when referring to previous publications of other geologists, and designations such as V series and T series are used herein when K-bentonites other than the Deicke and Millbrig are discussed.

The value of the Rocklandian K-bentonites in local correlations is established by the publications cited above, most of which hinted at (and several of which attempted to realize) the potential regional stratigraphic value of the K-bentonites as well. The regional correlations presented in this publication are in agreement with certain regional correlations in the Cincinnati Arch and westernmost Valley and Ridge that were based on chemical fingerprinting using both major and trace elements (Günal, 1979; Huff, 1983b), and geophysical logs (Huff and Kolata, 1990), and thus a unified regional stratigraphic framework is beginning to emerge. If the agreement between the regional correlations in the above publications and this publication is assumed to imply that large-scale regional stratigraphic relations are now correctly recognized, it can be shown that misinterpretation of regional K-bentonite stratigraphy has been the cause of significant stratigraphic confusion in some earlier papers either because: (1) certain assumptions were based on a particular correlation or group of correlations that cannot be correct in the context of current knowledge, (2) individual K-bentonites have not been identified in a consistent manner, or (3) the specific criteria on which identification of individual K-bentonite beds was based were not given.

For example, compare two papers published 15 yr apart that include a discussion by the same authors of the K-bentonites in the well-known Big Ridge exposure near Duck Springs, Alabama. Drahovzal and Neathery (1971) identified two thick K-bentonites there as being correlative with the K-bentonites T-3 and T-4 of Wilson (1949). They noted that, although both beds are relatively thick, it is only the T-4 K-bentonite that contains abundant biotite; this statement is in complete agreement with the findings of other geologists who have worked in the southeastern U.S., where the T series nomenclature of Wilson (1949) is used (Milici, 1969; Milici and Smith, 1969; Manley and Martin, 1972; Chowns and McKinney, 1980; Kunk and Sutter, 1984; Carter and Chowns, 1986). In the *Decade of North American Geology* (DNAG) volume, however, Neathery and Drahovzal (1986, p. 168) cited their earlier paper, stating that "The most distinctive bentonite bed is the biotite-rich T-3 bed (Drahovzal and Neathery, 1971)." This is confusing because the biotite-rich bed is unambiguously identified in the 1971 paper not as T-3, but as T-4, and also because the two K-bentonite beds identified as T-3 and T-4 in the Big Ridge stratigraphic column in 1986 are clearly *not* the same as the two identified as T-3 and T-4 in 1971. As discussed herein and elsewhere (Haynes and Huff, 1990), my work supports the correlations presented by Drahovzal and Neathery in the 1971 guidebook; unfortunately, the 1986 DNAG article will probably be read more widely than the guidebook.

Another example is found in the study of Ordovician conodonts by Hall (1986). The composite Chickamauga Limestone section of Hall (1986, p. 233–241) includes partial sections from several localities near Chickamauga, Georgia, including the section at Davis Crossroads (the same Davis Crossroads section of the present report) along the now-dismantled TA&G Railroad (Hall, 1986, p. 239–240). From a review of Hall's text and a comparison of his findings with those of Milici and Smith (1969) and those reported herein, it is apparent that the Carters Limestone to Catheys Limestone sequence has been misidentified by Hall, at least in part. First, the limestones in the Carters are repeatedly described as being coarse grained (Hall, 1986, p. 236–237), when in fact the Carters in the Chickamauga area and elsewhere in the southeastern United States is actually a very distinctive sequence of sparsely fossiliferous laminated lime mudstones (Milici, 1969; Milici and Smith, 1969). Second, Hall described a nearly complete section of the Catheys Formation at the Davis Crossroads section when it is actually the upper Lebanon, Carters, and lower Hermitage Formations that are exposed along the railroad grade in the low cut near Davis Crossroads (Milici and Smith, 1969). Hall described 6 in. of black chert in that section (about the same as the thickness of chert that I measured there beneath the Deicke), but there is no mention of K-bentonites, which are at this exposure in readily visible recessed zones. This apparent misidentification of stratigraphic units may have resulted in some erroneous conclusions by Hall concerning the age of some of the various units in the composite Chickamauga section, and thus in the correlation of strata based on conodonts that were obtained from those limestones.

An older but notoriously well-known example is the correlation of the Hounsfield metabentonite by Kay (1931) between many measured sections in the eastern midcontinent. A short time later Kay was forced to abandon this correlation (Kay, 1935) when it became clear that additional thick K-bentonite beds existed in many of the sections from which the Hounsfield was described, and Kay was subsequently unable to demonstrate a means of distinguishing one bed from another consistently and reliably.

These examples are three of several, and they point to the real need for a regional study of the Rocklandian K-bentonites in the southeastern U.S., especially the thickest, most widely persistent, and hence most stratigraphically useful ones. And they point to the need for a thorough investigation of some aspect of the petrology of each individual bed to document its uniqueness, ensuring the accuracy of the stratigraphic framework. Thus the purpose of the present publication, which attempts to address these needs, is threefold: (1) to demonstrate that Deicke and Millbrig samples from the Cincinnati Arch and the southern Valley and Ridge province, areas where these are demonstrably the thickest and most laterally persistent of the Rocklandian K-bentonites, are distinguishable in the field and the laboratory on the basis of nonclay mineralogy; (2) to discuss the diagenetic history of the Deicke and Millbrig and the Rocklandian sequence in the context of regional geologic history during the later Ordovician and the remainder of the Paleozoic; and (3) to clarify and expand the regional stratigraphy of the K-bentonite sequence, furthering the recent work of Huff and Kolata (1990) and Haynes (1992). Some reinterpretations of earlier correlations in the Valley and Ridge of Alabama and Georgia are presented herein; significant reinterpretations of earlier correlations in much of the Virginia Valley and Ridge were discussed by Haynes (1992), with only a synopsis of that work herein. Much of Haynes's work (1992) was based on the finding that the Deicke and Millbrig are distinguishable by their nonclay mineralogy, but the focus of that paper was primarily stratigraphic, hence a full discussion of the comprehensive mineralogic study behind the work was not included. Presented herein, however, is that discussion of the mineralogy, mineral abundances, composition, and stratigraphy of the Rocklandian K-bentonite beds in the southeastern United States, and how these findings have been used to develop a regional stratigraphic framework based on individual K-bentonite beds.

The present publication thus continues the study of Ordovician K-bentonites in the southeastern United States that began with their discovery and description by Nelson (1921, 1922, 1926); it is an expansion of, and a contribution to, the petrologic work of authors including Weaver (1953), Coker (1962), Huff (1963), Lounsbury and Melhorn (1964, 1971), Hergenroder (1966), Huff and Türkmenoğlu (1981), Kunk and Sutter (1984), Elliott and Aronson (1987), and Samson et al. (1989), and the stratigraphic work of Rosenkrans (1936), Fox and Grant (1944), Huffman (1945), Wilson (1949), Miller and Brosgé (1954), Miller and Fuller (1954), Milici (1969), Milici and Smith (1969), Drahovzal and Neathery (1971), Hergenroder (1973), Chowns and McKinney (1980), Huff (1983b), Kolata et al. (1984, 1986), Samson et al. (1988), Huff and Kolata (1990), Haynes (1992), and Huff et al. (1992).

METHODS

Elliott and Aronson (1987) showed that, unlike the Cretaceous bentonite studied by Altaner et al. (1984), the composition of the mixed-layer I/S in the Deicke and Millbrig (their T-3 and T-4 K-bentonites, respectively) in the southeastern U.S. is not vertically zoned. Therefore the minerals (clay and nonclay) in the coarsest grained zones in the Deicke and Millbrig were the primary focus because it was expected that detailed study of those phenocryst-rich zones would yield the most information on the origin and diagenesis of the two beds.

Samples were first collected along the Cincinnati Arch. They came from the outcrops in the Jessamine and Nashville Domes, and from the subcrop along the Cumberland Saddle between the two domes and from the subcrop in the Cumberland Plateau between the Cincinnati Arch and the Valley and Ridge. These samples were thin sectioned, polished, and then examined with the petrographic microscope to determine the composition of the two beds in that region and, to the extent possible, to confirm that those two beds were in fact the ones mentioned in many previous studies. Next, samples of K-bentonites collected from exposures in the southern Valley and Ridge province were studied in thin section and compared with the Cincinnati Arch samples. Again, samples from the coarsest zone or zones in each bed were examined. The <2-µm size fraction from samples from both the Cincinnati Arch and the Valley and Ridge was studied using x-ray diffraction. Once the general differences in phenocryst mineralogy between the Deicke and the Millbrig had been determined (Haynes et al., 1987), the regional variations in mineral compositions became the basis for an interpretation of the diagenetic history of the two beds throughout the southern Appalachian basin. This was done by examining distributions of pyrite versus hematite, ilmenite versus TiO_2 minerals, authigenic K-feldspar versus authigenic albite, and by studying the illite:smectite ratio and ordering of the mixed-layer I/S. The correlation of measured sections throughout the region was made possible by a combination of two factors: (1) the discovery that the two beds can be distinguished by their nonclay minerals, and (2) a detailed field study of the physical stratigraphy of the Middle and Upper Ordovician sequence.

Additional details of field sampling methods and sample preparation and analytical methods are in Appendix 1. Also in the appendix is a register of the localities (outcrops and boreholes) from which samples of the Deicke and Millbrig were obtained, and which are shown in Figure 1.

GEOLOGIC SETTING

Ordovician strata are present in surface exposures throughout much of the southeastern United States (Fig. 1). Many K-bentonites occur in these outcrops and in the adjacent areas of subcrop as well, and in several studies individual beds or groups of K-bentonite beds have been identified on the basis of thickness, texture, color, and the stratigraphic relations between K-bentonites and marker beds such as formation boundaries or faunal zones (Rosenkrans, 1933, 1936; Fox and Grant, 1944; Wilson, 1949; Miller and Brosgé, 1954; Miller and Fuller, 1954; Milici, 1969; Milici and Smith, 1969).

In their type areas in the Upper Mississippi Valley, the Deicke, Millbrig, and other Mohawkian K-bentonites are only a few centimeters thick and are noticeably lacking in nonclay minerals (Willman and Kolata, 1978; Kolata et al., 1986). Using geophysical logs of wells, Huff and Kolata (1990) correlated the Deicke and Millbrig from that region with the T-3 (Pencil Cave) and T-4 (Mud Cave) K-bentonites of the Cincinnati Arch and the western Valley and Ridge. The T-3 and T-4 beds are well known (Wilson, 1949; Milici, 1969; Milici and Smith, 1969; Drahovzal and Neathery, 1971; Chowns and McKinney, 1980), and are relatively thick and coarse grained. My initial examination of T-3 and T-4 samples from Cincinnati Arch cores immediately indicated that these coarse-grained samples would be ideally suited for a comprehensive petrologic study, which is the basis to understanding the origin of the volcanic material, and is the key to determining if the two beds can be distinguished on a basis other than geophysical character.

Figures 2 through 5 are photographs from throughout the study area of some of the best exposures of the K-bentonites now identified as the Deicke and Millbrig. Photographs of other excellent K-bentonite exposures in this region are found in Rosenkrans (1936), Butts (1940), Huffman (1945), Miller and Brosgé (1954), Miller and Fuller (1954), Huff and Kolata (1990), Haynes (1992), and in the open file of Rosenkrans's field work that is maintained at the Virginia Division of Mineral Resources, Charlottesville, Virginia (E. K. Rader, personal communication).

In these and other studies (Bay and Munyan, 1940; Fox and Grant, 1944; Wilson, 1949; Hergenroder, 1966; Milici, 1969; Milici and Smith, 1969; Drahovzal and Neathery, 1971), two K-bentonite beds were described as being much thicker and more laterally persistent than all other such beds in that region. In the upper bed, now recognized as the Millbrig, the presence of abundant biotite, particularly in a noticeably coarse-grained zone in the lower or middle part of the bed, was commonly reported. A coarse-grained zone was also reported from the lowest thick K-bentonite, which is now recognized as the Deicke, but that zone occurs in the basal several centimeters of the bed and it contains very little biotite. Each of the two K-bentonite beds is typically thicker than 30 cm, and at many exposures in the Valley and Ridge province each is 1 m or more thick, a

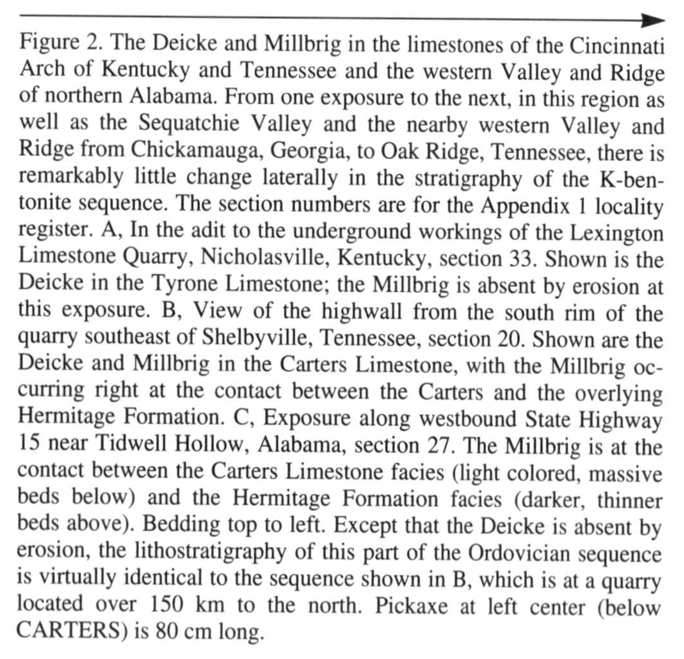

Figure 2. The Deicke and Millbrig in the limestones of the Cincinnati Arch of Kentucky and Tennessee and the western Valley and Ridge of northern Alabama. From one exposure to the next, in this region as well as the Sequatchie Valley and the nearby western Valley and Ridge from Chickamauga, Georgia, to Oak Ridge, Tennessee, there is remarkably little change laterally in the stratigraphy of the K-bentonite sequence. The section numbers are for the Appendix 1 locality register. A, In the adit to the underground workings of the Lexington Limestone Quarry, Nicholasville, Kentucky, section 33. Shown is the Deicke in the Tyrone Limestone; the Millbrig is absent by erosion at this exposure. B, View of the highwall from the south rim of the quarry southeast of Shelbyville, Tennessee, section 20. Shown are the Deicke and Millbrig in the Carters Limestone, with the Millbrig occurring right at the contact between the Carters and the overlying Hermitage Formation. C, Exposure along westbound State Highway 15 near Tidwell Hollow, Alabama, section 27. The Millbrig is at the contact between the Carters Limestone facies (light colored, massive beds below) and the Hermitage Formation facies (darker, thinner beds above). Bedding top to left. Except that the Deicke is absent by erosion, the lithostratigraphy of this part of the Ordovician sequence is virtually identical to the sequence shown in B, which is at a quarry located over 150 km to the north. Pickaxe at left center (below CARTERS) is 80 cm long.

striking contrast to the thickness of most of the other Rocklandian K-bentonites in the region.

In Mohawkian strata of the southern Appalachian basin, of which the Rocklandian K-bentonites are a part, carbonate rocks predominate in most of the area, with clastic rocks confined to the easternmost outcrop belts of the Valley and Ridge (Rodgers, 1953; Rader, 1982; Chowns and Carter, 1983a; Benson, 1986). The Deicke and Millbrig occur in the Tyrone Limestone of the uppermost High Bridge Group in central Kentucky; the Carters Limestone of the uppermost Stones River Group in central and southeast Tennessee and northwest Georgia; the upper Moccasin and lower Eggleston Formations in southwest Virginia, southeast West Virginia, and northeast Tennessee; and the upper Stones River Group (undivided) and Chickamauga Limestone in Alabama. The Stones River and Nashville Groups as defined and subdivided in the type area of the Chickamauga limestone (Milici and Smith, 1969) cannot be so divided in much of Alabama. Thus those names have been informally used at the formation level by Drahovzal and Neathery (1971), but they are referred to herein as Groups.

The clastic rocks along the eastern Valley and Ridge in which the K-bentonites occur are assigned to the Blount Group (Rodgers, 1953), a great clastic wedge that extends along strike from Alabama to Virginia but which persists laterally across only two or three thrust sheets in the southern Valley and Ridge. Both the Deicke and Millbrig occur in the Colvin Mountain Sandstone in Alabama, but only the Deicke has been found at sections of the Colvin Mountain Sandstone in Georgia and the Bays Formation in southeast Tennessee. In northeast Tennessee and southwest Virginia only the Millbrig is present in the Bays Formation.

Although Ordovician strata throughout the southern Valley and Ridge province are structurally disrupted to varying degrees, division into three stratigraphically distinct belts is possible in both Alabama and Georgia (Chowns, 1986) and in Virginia, West Virginia, and Tennessee (Kreisa, 1980; Rader, 1982; Haynes, 1992). The western belt consists of all exposures northwest of the Helena and Clinchport faults in Alabama and Georgia, and all exposures northwest of the Whiteoak Mountain, Hunter Valley, and St. Clair faults in Virginia and Tennessee. Rocklandian strata in this area are predominantly gray limestones of peritidal origin. The western boundary of the western belt is the eastern edge of the Appalachian Plateaus. The central belt includes exposures between the Helena and Eden faults (Alabama), the Clinchport and Rome faults (Georgia), the Whiteoak Mountain and Hunter Valley faults and the Saltville fault (Tennessee), the St. Clair and Saltville faults (southwestern Virginia), and the St. Clair and Pulaski faults (west-central Virginia). Rocklandian and associated strata in this belt are part of a sequence of nonmarine to transitional red and olive green calcareous siltstones and white quartzose sandstones. In the eastern belt of Alabama and Georgia, which includes exposures on all thrust sheets east of the Eden and Rome faults and west of the Talledaga and Cartersville faults, the youngest Ordovician strata are pre-Rocklandian based on conodont biostratigraphy (Hall et al., 1986). Those rocks are gray marine carbonates and calcareous black shales deposited in deep ramp to basin margin and basin settings. The eastern belt of Virginia and Tennessee includes exposures between the Saltville and Pulaski faults and the Blue Ridge (Great Smoky) fault. Here in this belt, unlike in Alabama and Georgia, a more nearly complete Ordovician section is present, and Rocklandian and associated strata are red calcareous sandstones and siltstones deposited in a tidal flat setting. These grade northward into noncalcareous greenish gray lithic siltstones and sandstones deposited in a delta front environment. South of Wytheville, Virginia, the outcrop belt east of the Pulaski fault and west of the Blue Ridge fault apparently does not include exposures of strata as young as Rocklandian (Rader, 1982), and no K-bentonites were found during a limited examination of the Knobs Formation, the youngest strata in this belt.

In the southern Valley and Ridge province, both the Deicke and Millbrig pass indiscriminately through the significant across-strike facies changes, as would be expected of beds of altered volcanic ash. This is true even though parallel outcrop belts tend to be on separate thrust sheets, and the facies changes are telescoped because of the overthrusting. Thus the Deicke and Millbrig are now recognizable in both carbonate and clastic strata of the Valley and Ridge province from Birmingham, Alabama, to Roanoke, Virginia.

PETROGRAPHY

Introduction

Petrographic study of thin sections and grain separates shows that the Deicke and the Millbrig K-bentonite Beds contain distinctly different nonclay mineral assemblages, and that the two beds can be distinguished from each other without ambiguity on this basis. The principal primary phenocrysts in the two beds are plagioclase feldspars (labradorite or andesine), quartz, biotite, and ilmenite, with minor zircon, apatite, and magnetite. The principal authigenic nonclay minerals in the two beds are K-feldspar, albite, TiO_2 (as rutile,, anatase, and/or brookite), hematite, and calcite, with minor chlorite and gypsum. Both beds contain mixed-layer I/S as the principal clay mineral, and kaolinite is present in some samples from the Colvin Mountain Sandstone and the Bays Formation.

Nonclay mineralogy

The clay mineralogy of Rocklandian K-bentonites in the southeastern U.S. has been discussed in many papers (Reade, 1959; Coker, 1962; Huff, 1963; Hergenroder, 1966; Lounsbury and Melhorn, 1964, 1971; Manley and Martin, 1972;

Millbrig
Deicke

S

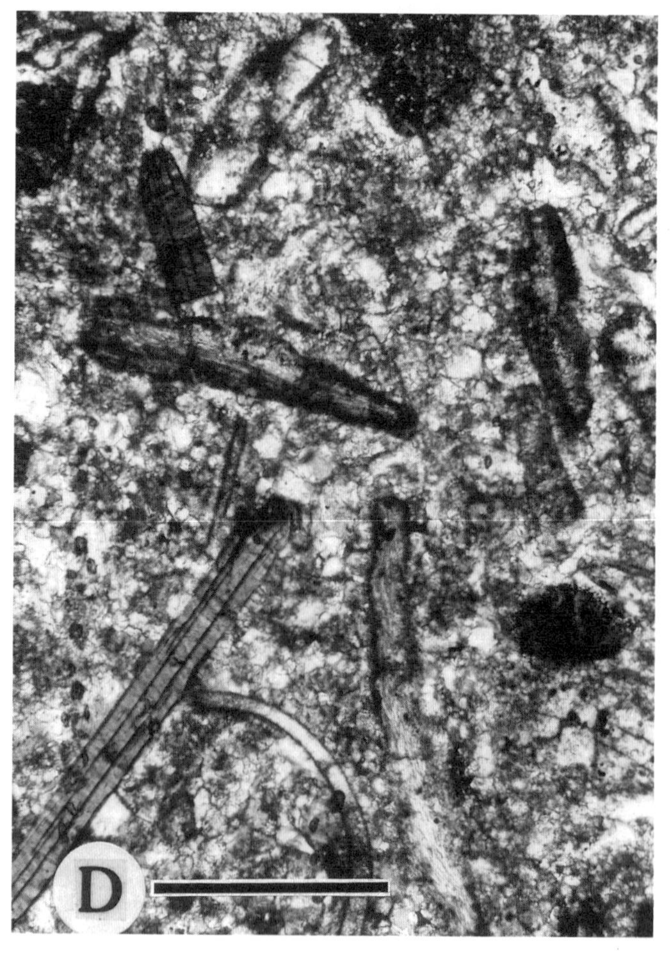

Huff and Türkmenoğlu, 1981; Elliott and Aronson, 1987), but comparatively little has been published on the nonclay minerals. Table 1 presents the abundances of the nonclay minerals in the Deicke and Millbrig as determined by the study of more than 80 thin sections with the petrographic microscope, and by analysis of 15 thin sections with the electron microprobe. Figure 6 shows the common nonclay minerals in the beds. Well-preserved primary phenocrysts are in general more abundant in the Millbrig than in the Deicke, presumably because the labradorite and Fe-Ti phenocrysts of the Deicke were more susceptible to alteration during diagenesis relative to the andesine, biotite, and quartz of the Millbrig.

No nonmagmatic primary components such as the xenocrysts and microliths reported from Ordovician K-bentonites in New York (Delano et al., 1990) were observed in any of the Deicke and Millbrig samples studied.

Feldspars. Feldspars in the Deicke and Millbrig are simply and polysynthetically twinned primary plagioclase, untwinned authigenic plagioclase, and untwinned authigenic K-feldspar (Fig. 6A, D, and Fig. 7). Untwinned K-feldspar is the most common feldspar, constituting up to 95% of all feldspars in some samples. Twinned or zoned plagioclases constitute up to 60% of the feldspars in some samples. Least abundant is untwinned albite. It is absent in samples from the Cincinnati Arch, but in some samples from the Valley and Ridge province, albite constitutes up to 30% of the total nonclay minerals. Primary potassium feldspars such as sanidine, which are common in felsic volcanic ash, are extremely rare. This absence is most likely a postdiagenetic texture; sanidine was probably dissolved and authigenic K-feldspar, which is abundant in both the Deicke and Millbrig, formed as a replacement.

Figure 3. The Deicke and Millbrig in the northwestern Valley and Ridge of Alabama; the section numbers are for the Appendix 1 locality register. A, Roadcut along I-59 northbound near Ft. Payne, section 24. Shown is the Deicke in fenestral micrites (Carters facies) of the Stones River Group. Both the Deicke and Millbrig form prominent recesses in the many excellent exposures along a several mile stretch of I-59 near Ft. Payne. B, Roadcut along I-59 northbound in the gap through Big Ridge, Alabama, section 25. Shown are the Deicke and Millbrig in the Stones River Group. C, Roadcut along the southbound Red Mountain Expressway, Birmingham, Alabama, section 29. Shown is the Millbrig in the Chickamauga Limestone. Bedding top to left. Reworking of original ash resulted in dispersal of bentonitic material throughout the interval between the arrows, making the Millbrig here more of a bentonite zone, which consists of the two discrete K-bentonite beds (arrows) and the intervening nodular shaly limestones (S) that contain abundant reworked bentonitic material mixed in with bioclastic debris. The Deicke occurs about 4 m downsection. D, Photomicrograph of biotite mixed in with bioclastic material in one of the calcareous nodules, a fossiliferous packstone, from the shaly zone shown in C. Scale bar is 0.5 mm long.

Grain size varies greatly. Average length of feldspars in samples from the Jessamine Dome of Kentucky and from the west-central Valley and Ridge in Virginia is less than 0.5 mm, whereas in samples from the Nashville Dome of Tennessee and from the Valley and Ridge of southwesternmost Virginia to Alabama, the average length is from 0.5 to 1.0 mm (Fig. 7). A detailed, quantitative study of grain size distribution for each zone of the K-bentonites is underway, and preliminary results show an overall increase in average grain size toward the south and southeast, Virginia to Georgia (Zhang and Huff, 1992). This thickening and coarsening is inferred to reflect increasing proximity to the source volcanoes.

The modal percentage of feldspars present varies with location as well. In the Jessamine Dome of Kentucky and the Valley and Ridge of west-central Virginia they average less than 30% of the nonclay minerals, whereas in samples from the Cumberland Saddle of Kentucky and Tennessee, the Nashville Dome of Tennessee, and the Valley and Ridge between Alabama and southwesternmost Virginia, they constitute as much as 60%. The major exception is samples from the eastern Valley and Ridge, where feldspars constitute from 0 to 40% of the nonclay minerals. In that area the beds are in the Blount Group, the clastic strata of which have been subjected to higher burial temperatures than the limestones farther west (Harris et al., 1978), resulting in more extensive alteration of the primary minerals.

The initial identification of feldspars was made with thin sections and the petrographic microscope. Many feldspars can be distinguished on a comparison of interference colors alone; in some samples euhedral zoned plagioclases are well preserved (Fig. 7A, B, E). Further identification of some grains was made using optical oils, but the most quantitative identification of feldspars was made by electron microprobe analyses. Twinned, zoned, and untwinned feldspars, and some grains in various stages of alteration, were all analyzed. Figure 8 summarizes the microprobe results in a ternary plot of data, and Table 2 lists representative analyses from several different feldspar grains. The presence of several different feldspars, indicated by the textural and semi-quantitative compositional observations made with the petrographic microscopic, is confirmed by the microprobe analyses.

The data points in the andesine and labradorite fields in Figure 8 were obtained exclusively from zoned or polysynthetically twinned plagioclases. All grains plotting in those fields have compositions close to the values for andesine and labradorite given in Table 2. Their compositions are typical of igneous plagioclases, which consist of solid solutions between the albite (Ab) and anorthite (An) end members, with a minor orthoclase (Or) component. Labradorite is defined as having 50 to 70% An component, and andesine as having 30 to 50% An component.

All data points obtained from analyses of untwinned or unzoned grains plotted in two corner fields (Fig. 8). Those

chert

fields represent the albite end-member composition in the Ab-An solid-solution series, and the orthoclase end-member composition in the Ab-Or solid solution series. All grains in the corner fields had compositions nearly identical to the values for albite and K-feldspar given in Table 2.

The apparent gap in miscibility seen between albite and orthoclase in Figure 8 actually reflects the selective diagenetic removal of nearly all the primary sanidine, of which only very few grains remain (R. L. Hay, personal communication). Miscibility gaps are characteristic of plutonic rocks formed deep in the crust from a slowly cooling melt, and it is assumed that the Deicke and Millbrig are not directly derived from such a melt (i.e., they are not sills), but instead are beds of altered volcanic ash. Thus the apparent gap is interpreted to reflect the various postdepositional changes that have affected the beds.

Most of the relatively unaltered twinned and zoned plagioclase grains in the Deicke are broken, but crystal faces can be seen along the unbroken edges and some euhedral grains are present (Fig. 7A, B, C). The twinned and zoned plagioclases in the Millbrig are texturally very similar to those in the Deicke.

Zoned and polysynthetically twinned plagioclase grains are characteristic of modern acid volcanic rocks. Low-temperature authigenic plagioclase that forms in sedimentary environments does not show zoning (Kastner and Siever, 1979), so the zoned plagioclases in the Deicke and Millbrig are interpreted to be primary phenocrysts. Further evidence supporting an igneous origin for the zoned and polysynthetically twinned plagioclases comes from their composition (Fig. 8). They are primarily andesines and labradorites, which are not known as authigenic phases (Kastner and Siever, 1979).

In both the Deicke and Millbrig, many of the twinned and zoned plagioclases have been partly or completely replaced by untwinned and stoichiometrically pure K-feldspar or albite (Figs. 6D and 7E). Based on their texture (Figs. 6 and 7) and composition (Fig. 8, endpoint data), these are interpreted to be authigenic feldspars. Feldspar authigenesis processes have produced several petrographically distinct "two-feldspar" grains, in which moderately to highly altered primary twinned or zoned plagioclases coexist with untwinned and less altered authigenic albite or authigenic K-feldspar. Replacement commonly occurs along planes of twinning or zoning (Figs. 6A and 7E). In addition, authigenic albite and authigenic K-feldspar also coexist in many grains (Fig. 7D), with albite commonly occurring around the edges of the grains.

Authigenic albite and authigenic K-feldspar also occur as individual grains, and together in single grains as another form of two-feldspar grains (Fig. 7A, C). Some of these grains are undoubtedly complete replacements after a primary plagioclase or sanidine.

Albitized K-feldspar grains with textures very similar to that in Figure 7D are illustrated and described by Morad (1988a) from Upper Proterozoic sandstones of Norway. In those sandstones, detrital microclines have been albitized, and the evidence indicates that the alteration is postdepositional rather than an inherited texture from albitized source rocks. With the Deicke and Millbrig, however, there is no doubt that the albitization was postdepositional. The K-bentonite beds are derived from potassium-rich volcanic ash, and the replacement of primary plagioclase by authigenic albite must have occurred subsequent to the deposition of the ash because magmatic replacement would have resulted in a phase that was also potassium-rich, which albite is not. Also, the plagioclase feldspars in the K-bentonites are not detrital, as they would be in a sandstone, so there is no possibility that the observed textures were inherited from older rocks.

Under cross-polarized light, authigenic K-feldspars have a patchy to ragged or tattered extinction (Fig. 7C). This readily distinguishes them from the authigenic albites, which have a much more uniform extinction. The authigenic K-feldspar was not studied further to determine its precise crystallographic identity.

Figure 4. The Deicke and Millbrig in the Valley and Ridge of Alabama, Georgia, and Virginia; the section numbers are for the Appendix 1 locality register. A, Exposure along the former railroad right-of-way near Davis Crossroads, Georgia, section 22. Shown is the almost 1.5-m-thick Deicke in the Carters Limestone. The lower contact is the top of the underlying chert layer, indicated by the arrows; the upper contact, not visible, is in the slope above the 80-cm-long mattock handle. The Millbrig is about 9 m upsection. B, Roadcut along State Highway 77 northbound through Greensport Gap, Alabama, section 48. Shown is the 8- to 10-cm-thick Deicke in the Colvin Mountain Sandstone. The arrow points to the top of the Deicke proper, but there is appreciable bentonitic material in the matrix of the overlying white sands, which are mature cross-bedded quartz arenites. Hammer is 26 cm long. C, Exposure along the railroad at Daleville, Virginia, section 16. Shown is the Millbrig in sandstones and siltstones of the Bays Formation, bedding top to right. Because of the near vertical attitude of the strata, the Millbrig has weathered into a recessed interval (arrow) that rapidly fills with wash from the vegetated slope above. The Deicke is absent by erosion, but K-bentonite bed V-7 occurs about 18 m upsection. D, Excavation in nearly horizontal strata along an anticlinal axis near Chickamauga, Georgia, section 52. Shown is the Deicke in the Carters Limestone. The base of the Deicke is the chert layer, and the top is shown by the arrow.

Deicke
chert
A

Deicke
B

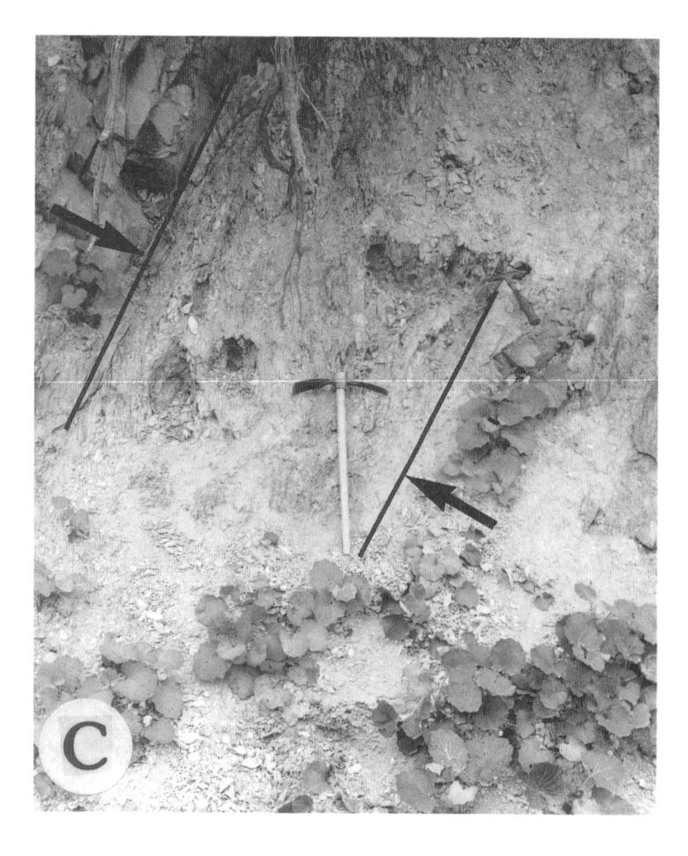
C

16
TAZEWELL 24
BISHOP 35
D

Two-feldspar albite and K-feldspar grains do not occur in samples from the Cincinnati Arch because albite is completely absent from that area. K-feldspars therefore constitute up to 85% of the feldspar population in Cincinnati Arch samples, with twinned or zoned primary plagioclase making up the remainder. In samples from the western Valley and Ridge, however, albite is common, and up to 25% of the feldspars are two-feldspar grains, containing both authigenic K-feldspar and albite, or one or both of those feldspars as overgrowths or replacements around the edges of primary plagioclases.

It is evident from Figure 8 that the Deicke and Millbrig can be distinguished by the composition of their most abundant primary plagioclase phenocrysts. Labradorite occurs in the Deicke, whereas andesine occurs in the Millbrig, and the data plot in distinctly separate fields. In any igneous rock the composition of primary feldspars is very useful in drawing inferences about the composition of the original melt, and the presence of zoned labradorites in the Deicke suggests that the original magma was of an intermediate composition, whereas the Millbrig magma, which contained andesine, is inferred to have been more felsic in composition.

In contrast to the identifiable difference between the primary plagioclases of the two beds is the abundance of end-member albite and K-feldspar in both beds. Composition of those feldspars does not vary significantly (Fig. 8, endpoint data), and given the authigenic origin of the albite and K-feldspar, no significant variation in composition would be expected. Thus the presence of those feldspars is not diagnostic of either the Deicke or the Millbrig. The K-feldspars are considered to be authigenic based on their textural relations with the primary labradorites and andesines, their lack of zoning or polysynthetic twinning, and their compositional purity, a distinguishing trait of feldspars formed in low-temperature (less than 400°C), aqueous sedimentary environments (Kastner and Siever, 1979).

In summary, feldspars are the most abundant nonclay minerals in samples of the Deicke and Millbrig from the limestones of the Cincinnati Arch and western Valley and Ridge. The apparent lack of feldspars in the few samples studied by Samson et al. (1988) from the Carters Limestone and Stones River Group of the Cincinnati Arch and western Valley and Ridge is a finding that should thus be viewed with some caution, considering that feldspars are ubiquitous in Deicke and Millbrig samples from the limestones of that region.

Fe and Fe-Ti minerals. The Deicke contains various nonsilicate ferroan and ferrotitanian minerals in abundances ranging from 5 to 10% of the nonclay minerals in most sam-

TABLE 1. FRAMEWORK GRAINS IN THE DEICKE AND MILLBRIG K-BENTONITES*

Millbrig		Deicke	
PRIMARY PHENOCRYSTS			
	(%)		(%)
Andesine	(1–15)	Labradorite	(1–15)
Quartz	(20–45)	Ilmenite	(<1–5)
Biotite	(15–50)	Magnetite	(<1)
Apatite	(<1)	Quartz	(<1)
Zircon	(<1)	Biotite	(<1)
		Apatite	(<1)
		Zircon	(<1)
AUTHIGENIC MINERALS			
K-feldspar - Or_{90-100}	(5–60)	K-feldspar - Or_{90-100}	(5–80)
Albite-Ab_{0-20}	(0–20)	Albite-Ab_{90-100}	(0–20)
Pyrite	(<1)	TiO_2†	(0–10)
Hematite	(<1)	Hematite	(<1–5)
Chlorite	(<1)	Pyrite	(<1–5)
Calcite	(<1)	Calcite	(<1–5)
		Gypsum	(<1)

*Albite is present in samples from the Valley and Ridge only. Abundances are expressed as a percentage of the total volume of nonclay minerals in the K-bentonites as determined from study of 81 thin sections.

†As anatase or rutile, and possibly brookite. Some leucoxene is probably also present.

Figure 5. The Deicke and Millbrig in the Valley and Ridge of Virginia; the section numbers are for the Appendix 1 locality register. A, Outcrop along the rail switchback at Hagan, section 1. Shown is the Deicke in the Eggleston Formation. Bedding top to left. The thick (3-6 cm) chert layer immediately beneath the recessively weathered K-bentonite forms a prominent ledge in the exposure. The Millbrig occurs about 12 m upsection. Photo by W. D. Huff. B, Exposure at Gate City, section 2. Shown is the Deicke in the upper Moccasin Formation redbeds. Bedding top to left. The lower contact is the light colored bed (white line), also visible in the foreground; the upper contact is the dashed line. Where it occurs in the Moccasin redbeds, as here, the Deicke is also noticeably red. C, Exposure at Gate City, section 2. Shown is the Millbrig in the thin Eggleston Formation upsection from the Deicke in B, bedding top to left. The actual thickness of the Millbrig is shown, with the arrows and lines indicating the lower and upper contacts, but the true thickness is less, because the Eggleston is severely disrupted structurally and the Millbrig has been thickened by duplication of bedding. The mattock is 80 cm long. D, Roadcut along northbound State Highway 16 south of Chatham Hill, section 10. Shown is the Millbrig in the red siltstones and fine-grained sandstones of the Bays Formation. Bedding top to right. Arrows and white line indicate the lower and upper contacts. The Walker Mountain Sandstone Member crops out 28 m downsection, at the interval where the Deicke is expected to occur. Although there is complete exposure here, the Deicke is absent, as it is in all sections of the Bays Formation in northeasternmost Tennessee and Virginia.

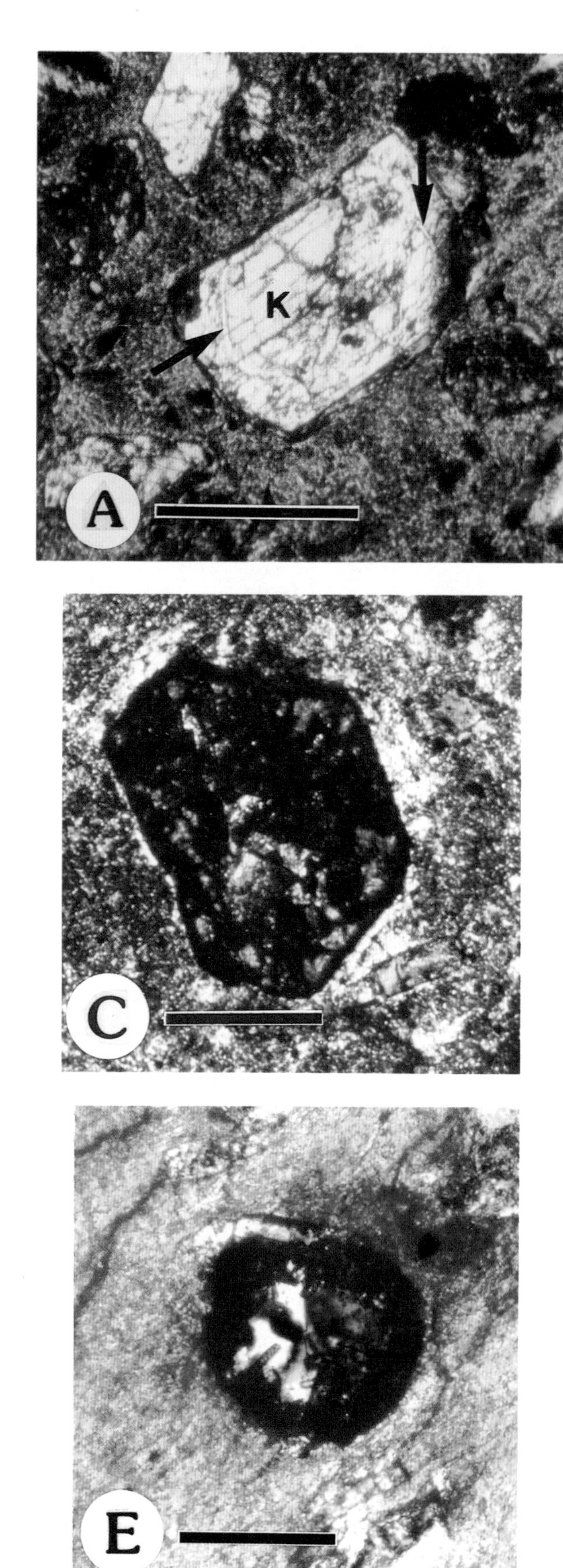
K
A
C
E

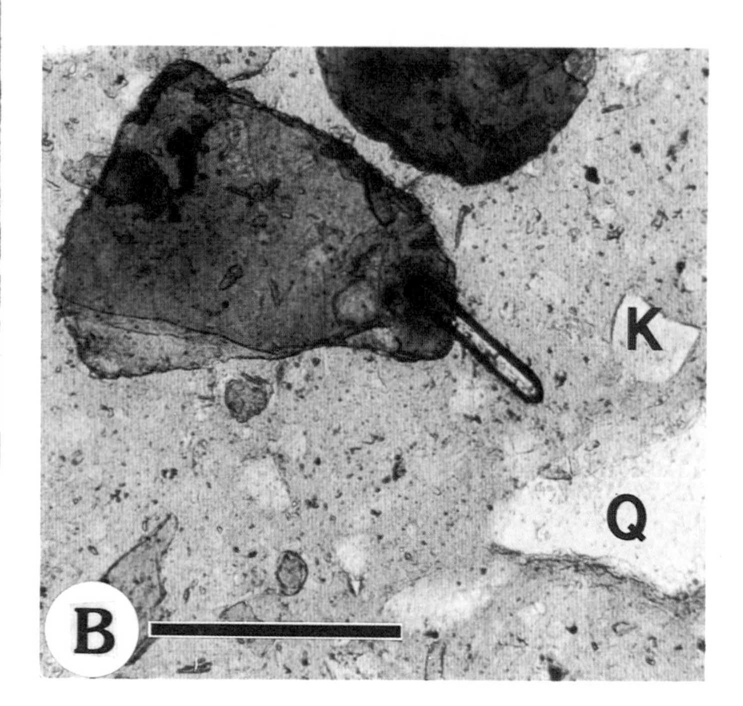
K
Q
B

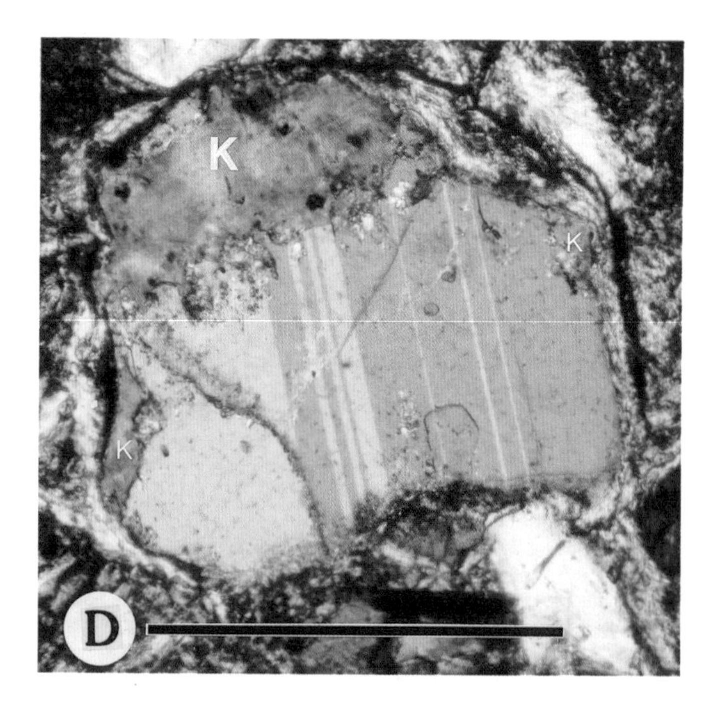
K
K
K
D

Figure 6. Photomicrographs of the common nonclay minerals in the Deicke and Millbrig. The matrix is mixed-layer illite/smectite in all samples. Scale bar is 0.5 mm long except as noted. The section numbers are for the Appendix I locality register. A, Euhedral plagioclase grain completely replaced by authigenic feldspars. Relict zoning is preserved in the secondary albite (arrows), and the presence of K-feldspar (K) in the otherwise albitized grain is evident from the differences in interference colors. From the Deicke, Carters Limestone, Davis Crossroads, Georgia, section 22. Cross-polarized light. B, Biotites, quartz (Q), and K-feldspars (K), with a zircon crystal protruding from the large biotite grain. From the Millbrig, Tyrone Limestone core, Marion County, Kentucky, section 41. Plane-polarized light. C, Euhedral ilmenite grain completely pseudomorphed by TiO_2. The high-order interference colors typical of the anatase and/or rutile distinguish these grains from all others. Scale bar is 0.2 mm long. From the Deicke, Carters Limestone, Trenton, Georgia, section 50. Cross-polarized light. D, Polysynthetically twinned andesine grain, average composition An_{38}, that is partly replaced by authigenic K-feldspar (K) in which the twin lamellae are not preserved. Compare with the plagioclase grain in A, in which relict zoning is preserved by the authigenic albite. From the Millbrig, Stones River Group, Big Ridge, Alabama, section 25. Cross-polarized light. E, Exceptionally large TiO_2 crystals in a void previously occupied by the precursor ilmenite grain. All pseudomorphed ilmenite grains contain clusters of these small acicular TiO_2 crystals, readily identifiable by their high-order interference colors but with crystal faces only rarely visible in thin section. Section is thicker than normal. Scale bar is 0.2 mm long. From the Deicke, Carters Limestone core, Clay County, Tennessee, section 45. Cross-polarized light.

ples. In the Millbrig, pyrite is the only nonsilicate ferroan mineral present, constituting less than 1% of the nonclay minerals, and it is probably secondary. Examination of polished thin sections of Deicke samples in reflected light shows that pyrite is common, and magnetite/titanomagnetite and ilmenite are less common. TiO_2 minerals, recognizable in transmitted light (Fig. 6C), are very common pseudomorphs of ilmenite. Figure 9A and B shows the euhedral outline of original ilmenite grains that have been completely replaced by TiO_2 minerals.

From his work in Tennessee, Coker (1962) reported that anatase was present in an Ordovician K-bentonite that is most likely the Deicke (Haynes, 1992). Johnsson (1986) reported anatase in the <2-μm size fraction of Tioga K-bentonite samples (Devonian) from upstate New York, and Lyons et al. (1992b) reported ilmenite, rutile, anatase, and brookite from the Fire Clay tonstein (Pennsylvanian) of the Appalachian basin. Studies of detrital titanomagnetite grains in sandstones indicate that both anatase and rutile can pseudomorph the ilmenite lamellae in such grains (Morad, 1988b). Together, these findings suggest that TiO_2 is not an uncommon secondary mineral in some Paleozoic K-bentonites and tonsteins.

As discussed by Morad and AlDahan (1986), authigenic TiO_2 comes in many colors and shapes, and various intermediate products occur during the alteration of ilmenite, with each phase successively enriched in titanium and depleted in iron. The TiO_2 minerals in the Deicke are identified on the basis of habit, red to brown color, high-order interference colors, and association with pseudomorphed euhedral grains of ilmenite (Figs. 6C and 9A, B). Further study of the TiO_2 minerals is underway, but at present no definite identification of the polymorph(s) present in the beds has been made. The TiO_2 minerals are commonly various shades of reddish brown in polarized light. Under high magnification the individual euhedral to subhedral highly birefringent crystals of TiO_2 are visible in sagenite clusters (Fig. 6E).

In one sample TiO_2 minerals are pseudomorphic after a grain whose euhedral outline is suggestive of a pyroxene rather than ilmenite (Fig. 9C). Pyroxene would not be unexpected in an ash of intermediate composition, so it is possible that the TiO_2 minerals in the Deicke are replacing both ilmenite and pyroxene phenocrysts. Pyroxene phenocrysts were not observed in the Deicke, however.

Samples of the Deicke from the Colvin Mountain Sandstone in Alabama and Georgia contain euhedral black ilmenite, which is unaltered or slightly altered (to leucoxene). The ilmenite is identifiable in thin section as opaque polygonal grains (Fig. 9D) that are whitish gray in reflected light. In grain separates the hexagonal shape and numerous crystal faces of the small, shiny black grains are readily identifiable under the binocular microscope.

Authigenic pyrite in the Deicke occurs in several shapes including small grains 0.1 to 0.3 mm long, and larger, irregularly shaped void fillings that have an intricate boxwork appearance in polished samples. The small isolated grains commonly are cubic (square in thin section); they may be pseudomorphic after magnetite. In the Millbrig, authigenic pyrite occurs in biotite grains as small cubes less than 0.1 mm long that are commonly sandwiched by individual sheets within the biotite grain.

Samples of the Deicke from the Moccasin Formation in Virginia, West Virginia, and Tennessee are predominantly red, with some yellow-brown zones. The red color is from oxidized iron that is disseminated throughout the bed via adsorption onto the I/S. In thin section, hematite is also visible as small blebs and patches that are opaque except around the edges, where the deep red color can be seen. No pyrite, TiO_2 minerals, or ilmenite were observed in Deicke samples from the Moccasin Formation.

Quartz. All Millbrig samples contain abundant quartz phenocrysts that are the same size as the feldspars (Fig. 10A), and quartz averages 25% of the nonclay minerals. By comparison, the Deicke samples contain less than 1% quartz phenocrysts. Quartz grains are subhedral to anhedral and are unaltered, textures that aid in distinguishing quartz from K-feldspar and untwinned plagioclase. There is no evidence of strain (undulatory extinction), and no overgrowths were seen.

Biotite. Abundant coarse euhedral to anhedral biotite flakes up to 1.5 mm long are the most obvious mineral in hand specimens of the Millbrig (Figs. 6B and 10B, C). Biotite makes up approximately 30% of the nonclay minerals in most

A

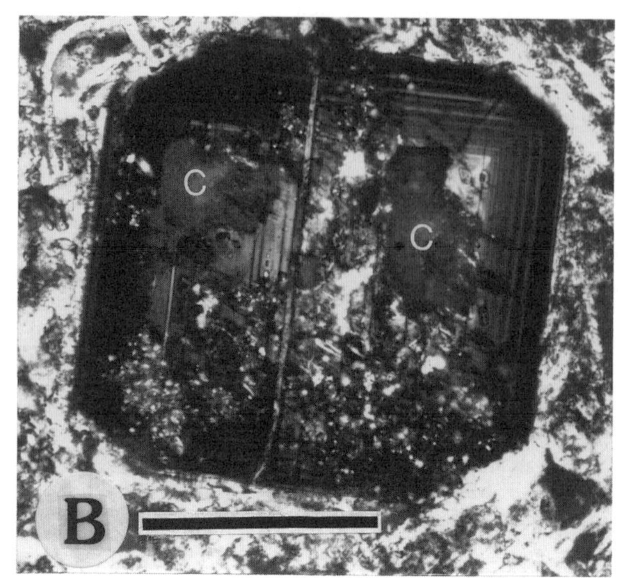
C
C
B

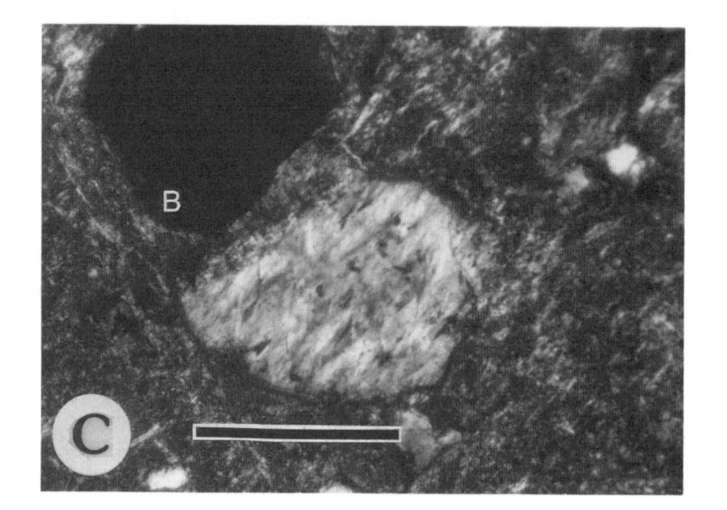
B
C

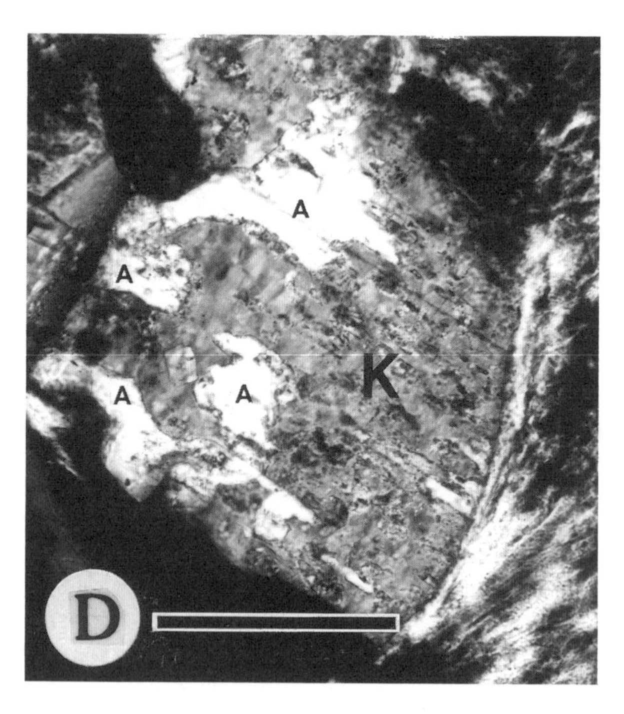
A
A
A
A
K
D

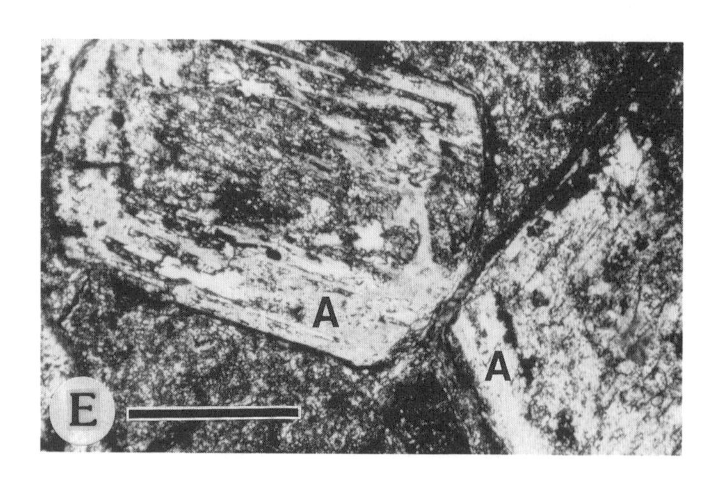
A
A
E

Figure 7. Feldspars in the Deicke and Millbrig. The matrix is mixed-layer illite/smectite in all samples. Scale bar is 0.5 mm long. The section numbers are for the Appendix 1 locality register. A, Euhedral plagioclases that appear to be "floating" in the matrix, a typical texture of feldspars in porphyritic volcanic rocks. From the Deicke, Carters Limestone, Rising Fawn, Georgia, section 23. Plane-polarized light. B, Zoned euhedral labradorite grain, average composition An_{57}. Secondary calcite (C) occurs in the core of the grain. From the Deicke, Chickamauga (Carters) Limestone or Eggleston Formation, Clinton, Tennessee, section 18. Cross-polarized light. C, Typical of authigenic K-feldspars in the Deicke and Millbrig, a K-feldspar grain with optically patchy or ragged extinction. Adjacent grain is biotite (B). From the Millbrig, Tyrone Limestone, Carntown, Kentucky, section 34. Cross-polarized light. D, Authigenic albite in a grain of authigenic K-feldspar. The higher order (white) interference color of the optically clear albite (A) contrasts with the lower order (gray) color of the K-feldspar (K), which has the optically ragged texture characteristic of those grains. From the Millbrig, Carters Limestone core, Marshall County, Alabama, section 47. Cross-polarized light. E, Primary labradorite grains that have been extensively albitized (A). Note the slightly higher order interference color of the albite relative to the labradorite, and the relict zoning fabric. From the Deicke, Carters Limestone, Davis Crossroads, Georgia, section 22. Cross-polarized light.

Millbrig samples, and less than 1% in the Deicke. Biotite occurs both as small slivers and larger flakes. The largest primary or secondary nonclay minerals in all K-bentonite samples are biotite grains, and it is the presence of a coarse-grained, biotite-rich zone near the lower part of the Millbrig that distinguishes it in the field.

In core samples and in samples from less weathered outcrops the biotite is very dark brown to black and at a glance appears unaltered. In thin section, biotite grains from those samples are usually pleochroic and range in color from pale green to deep brown, but most are red-brown (Fig. 6B). Even in biotites that appear virtually unaltered, however, some chloritization has occurred, as scanning transmission electron microscope (STEM) studies of biotites from the Millbrig show that some degree of chloritization is ubiquitous in nearly all grains (M. J. Kunk, personal communication). Samples collected from outcrops tend to be more weathered than those from cores, and in outcrop samples the biotites commonly are a deep golden or bronze color rather than dark brown or black. In sections cut parallel to bedding, the hexagonal outline common to basal sections of biotite is evident in many grains (Fig. 10C).

Biotites in Millbrig samples from some localities in the Bays Formation in Virginia and Tennessee are a pale green, indicating that chloritization has been more extensive. Many of those grains also exhibit a vermiform texture, suggesting that partial alteration to vermiculite and possibly kaolinite has occurred along individual sheets within a single biotite grain. Even with this expanded, vermiform texture, however, they are clearly identifiable as altered biotites (Fig. 10D).

In nearly all samples, the biotite grains, whether chloritized or not, are aligned parallel to bedding. This alignment reflects both depositional fabric and a reorientation strain fabric that developed normal to the stresses that affected the ash beds during compaction and alteration of the glassy matrix. Many biotites are squeezed around feldspar and quartz grains, a compaction texture that probably developed during post-depositional burial of the sediments. Because of their platy habit, the biotites deformed around the harder grains. No significant alignment of biotites parallel to axial plane cleavages was seen in Millbrig samples from the Valley and Ridge.

Several dark brown to black biotite grains were selected for microprobe analyses. Table 3 presents analyses of biotite grains from four samples of the Millbrig. The samples were not analyzed for OH^- or F^-. One of the analyses in Table 3 is of a more highly altered biotite grain from the Millbrig at Ft. Payne, Alabama (AL 14-1). In hand sample, the biotites from the Ft. Payne sample are brown in color, but in thin section a greenish tint is visible along many of the cleavage planes, a texture sug-

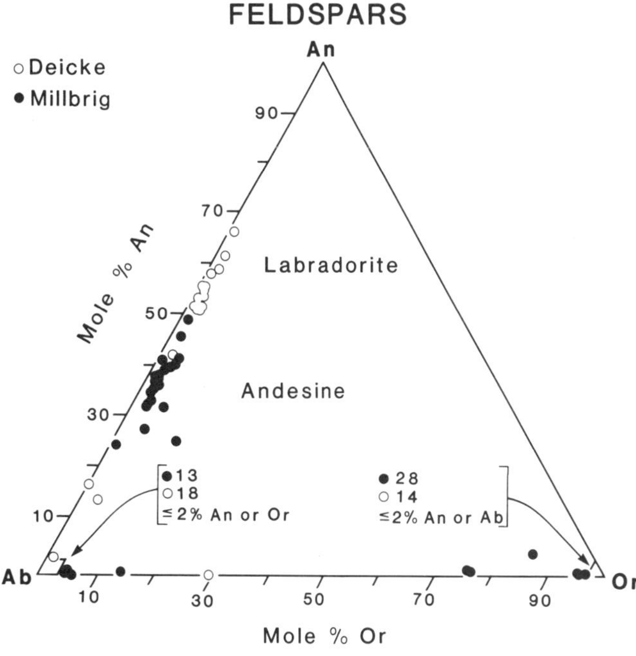

Figure 8. Results of microprobe analysis of feldspars in 15 thin sections of selected Deicke and Millbrig samples from Kentucky, Virginia, Tennessee, Georgia, and Alabama. Each data point represents a single feldspar grain analysis; no analytical traverses of zoned plagioclases were made because of the rarity of grains suitable for such analysis. The primary plagioclase data points separate into a labradorite field (Deicke) and an andesine field (Millbrig). These data points are all from the analysis of zoned or twinned plagioclases. The feldspars of extreme stoichiometric purity, which plot in the corner fields and are indicated by the arrows, are the untwinned and unzoned authigenic albite and K-feldspar grains that are ubiquitous in most samples. The abundance of these stoichiometric feldspars is indicated by the number of samples that plotted in the endmember Or and Ab fields.

TABLE 2. CHEMICAL ANALYSES OF FELDSPARS*

	Millbrig					Deicke				
Sample	AL 17	AL 17	KB-34	AL 14-1	TN 11-1	GA:WL 3	GA:WL 3-1	TN 10	KB-35	TN 12
Volumes in wt. %										
SiO_2	59.88	64.12	64.63	71.07	57.41	55.12	68.26	54.85	65.49	66.28
Al_2O_3	25.09	18.77	19.28	19.31	26.43	28.07	19.76	28.78	18.58	18.30
CaO	7.59	0.00	0.05	0.00	9.32	11.35	0.15	12.13	0.07	0.13
Na_2O	7.24	0.20	0.13	11.11	5.98	5.13	11.09	3.97	0.11	0.10
K_2O	0.71	16.84	16.16	0.00	0.38	0.45	0.23	0.3C	15.13	15.49
Total	100.51	99.93	100.25	101.49	99.52	100.12	99.49	100.03	99.38	100.30
Ab	60.9	1.8	1.1	100.0	52.5	43.8	98.2	36.6	1.0	0.7
Or	3.8	98.2	98.8	0.0	2.3	2.5	1.2	1.9	98.8	98.8
An	35.3	0.0	0.1	0.0	45.2	53.7	0.6	61.5	0.2	0.5
Grain type	Andesine	K-spar	K-spar	Albite	Andesine	Labradorite	Albite	Labradorite	K-spar	K-spar

*Sample localities are given in Appendix 1.

gesting that chloritization has begun. Compared with the other analyses in Table 3, it is evident that the amounts of K, Mn, Fe, and Ti in the Ft. Payne sample have decreased, and the amounts of Mg, Na, and Ca have increased, changes that commonly occur when biotites are chloritized. These trends support the petrographic observations, which indicate that the Ft. Payne biotites have been partly chloritized, because Ti and Mn occur only as trace elements in octahedral coordination in chlorites, a contrast to their often significant octahedral component in biotites. As shown in Table 3, the marked decrease in Mn that occurs during chloritization is especially apparent in the Ft. Payne sample, as it no longer even contains any detectable Mn. Therefore the original biotites have unquestionably been chloritized to some extent, but because significant amounts of K and Ti remain in the crystal structure, they are still biotites (albeit green biotites, an intermediate stage in the chloritization process), rather than true chlorites. The absence of any Ti-bearing authigenic minerals and the relative scarcity of pyrite and calcite in Millbrig samples from all localities, including Ft. Payne, further indicate that Ti and Fe are still a significant part of the biotite crystal structure.

The composition of chlorites and chloritized biotites from the Millbrig in the Bays Formation was not determined because of the generally inferior quality of those particular thin sections and because the chlorites and altered biotites in samples of the Bays failed to take a polish suitable for electron microprobe analysis. The noticeable green color of those grains in thin section and hand sample, however, suggests that chloritization has been more extensive in this area than farther west.

Analysis of the variation in the mean weight percents of FeO/MgO and TiO_2 in biotites is a demonstrated means of distinguishing thin beds of tuff and volcanic ash in sedimentary sequences (Desborough et al., 1973). Biotites in several Millbrig samples were analyzed by microprobe, and the mean value of the data was plotted using the technique of Desborough et al. (1973) (Fig. 11). For comparison, the mean values of samples from five of the tuff beds studied by Desborough et al. (1973) are also plotted in Figure 11. These data suggest that the Millbrig biotites may in fact be characterized by unique proportions of titanium, iron, and magnesium, but additional samples will be studied to improve the data set and increase the confidence level of these preliminary findings, as well as to test the correlation method of Yen and Goodwin (1976), who reported that correlation using biotite compositions is improved when three additional elements (Mn, Ni, Cr) are used in addition to Fe, Mg, and Ti, and a discriminant analysis program is employed.

The use of compositional variation among individual volcanogenic minerals in K-bentonites (apatite, zircon, etc.) is potentially a very useful correlation tool (Samson et al., 1988). Although whole-rock trace element chemistry is now widely used to correlate Cenozoic volcanic ash beds (Borchardt et al., 1971, 1973; Westgate and Fulton, 1975) and Paleozoic K-bentonites (Huff, 1983b; Cullen-Lollis and Huff, 1986; Kolata et al., 1986), that method appears best suited for comparing samples with similar (by volume) phenocryst/phenoclast-to-clay ratios (W. D. Huff, personal communication). This drawback is one of the likely reasons that attempts to correlate the Deicke and Millbrig from the Upper Mississippi Valley to the Cincinnati Arch based on chemical fingerprinting have met with limited success, forcing the use in that area of other correlation tools such as phenocryst assemblages (present study) and wireline logs (Huff and Kolata, 1990). Compositional variations in titanium, iron, and magnesium in Millbrig biotites, when plotted as a group centroid (Fig. 11), may eventually prove to be an additional means of demonstrating correlation of the K-bentonites, as might the correlation technique of Yen and Goodwin (1976).

As additional Deicke samples are collected, it is antici-

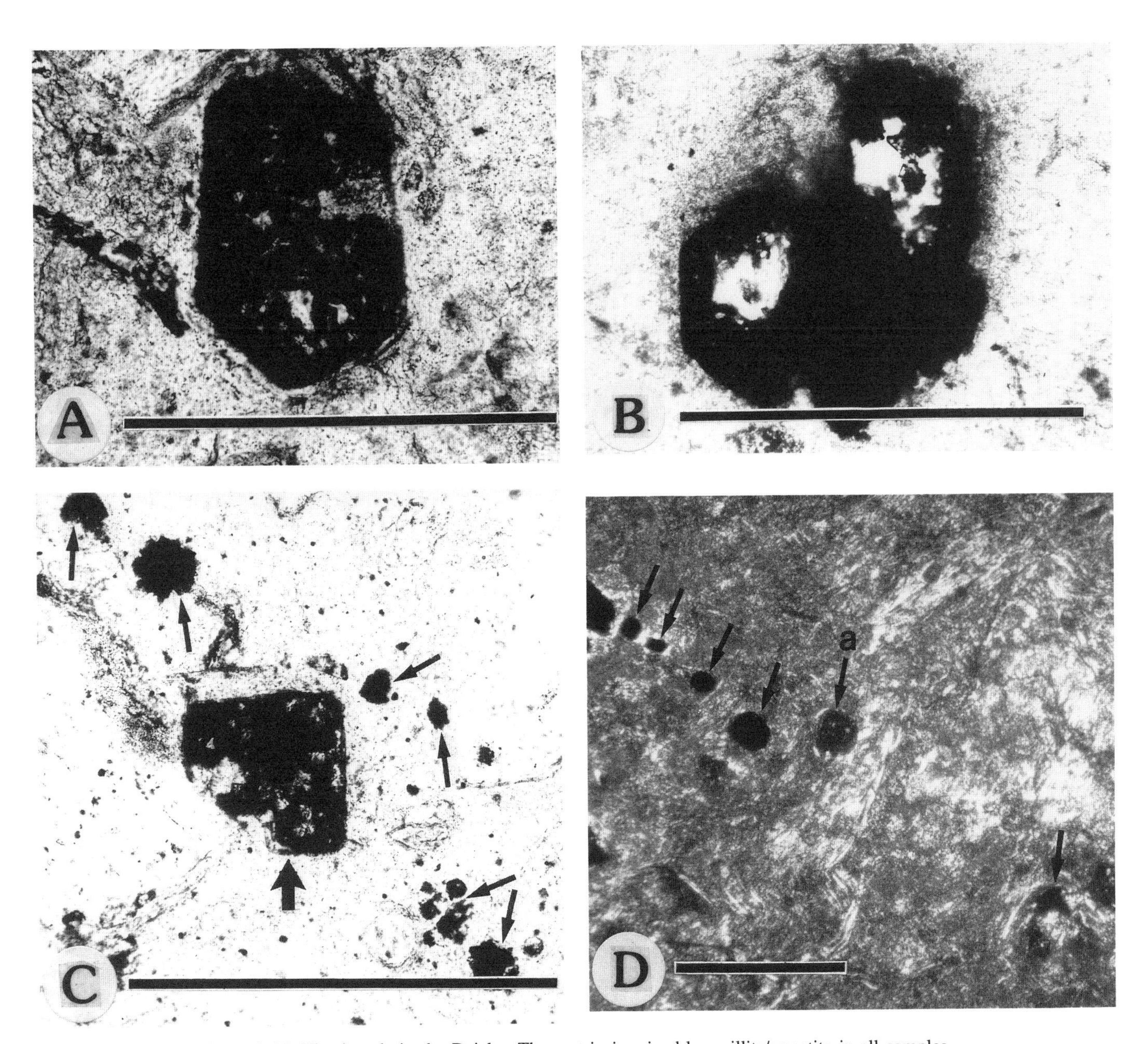

Figure 9. Fe-Ti minerals in the Deicke. The matrix is mixed-layer illite/smectite in all samples. Scale bar is 0.5 mm long. The section numbers are for the Appendix 1 locality register. A, Euhedral ilmenite grain completely pseudomorphed by TiO_2. From the Deicke, Carters Limestone, Trenton, Georgia, section 50. Plane-polarized light. B, A cluster of TiO_2 crystals replacing three separate euhedral ilmenite grains. The voids within the pseudomorphed ilmenite are a common feature of these grains. From the Deicke, Tyrone Limestone core, Casey County, Kentucky, section 42. Plane-polarized light. C, TiO_2 crystals (large arrow) pseudomorphous after what was most likely a pyroxene grain, whose euhedral outline is readily apparent. Pyrite (small arrows) is also common in many Deicke samples. From the Deicke, Carters Limestone core, Clay County, Tennessee, section 45. Plane-polarized light. D, Ilmenite grains. The four subhedral to euhedral opaque grains at left (arrows) are little altered, while the translucent grain at right (arrow marked a) has been partly altered to TiO_2 and/or leucoxene. There is appreciable kaolinite in the matrix. From the Deicke, Colvin Mountain Sandstone, Alexander Gap, Alabama, section 28. Plane-polarized light.

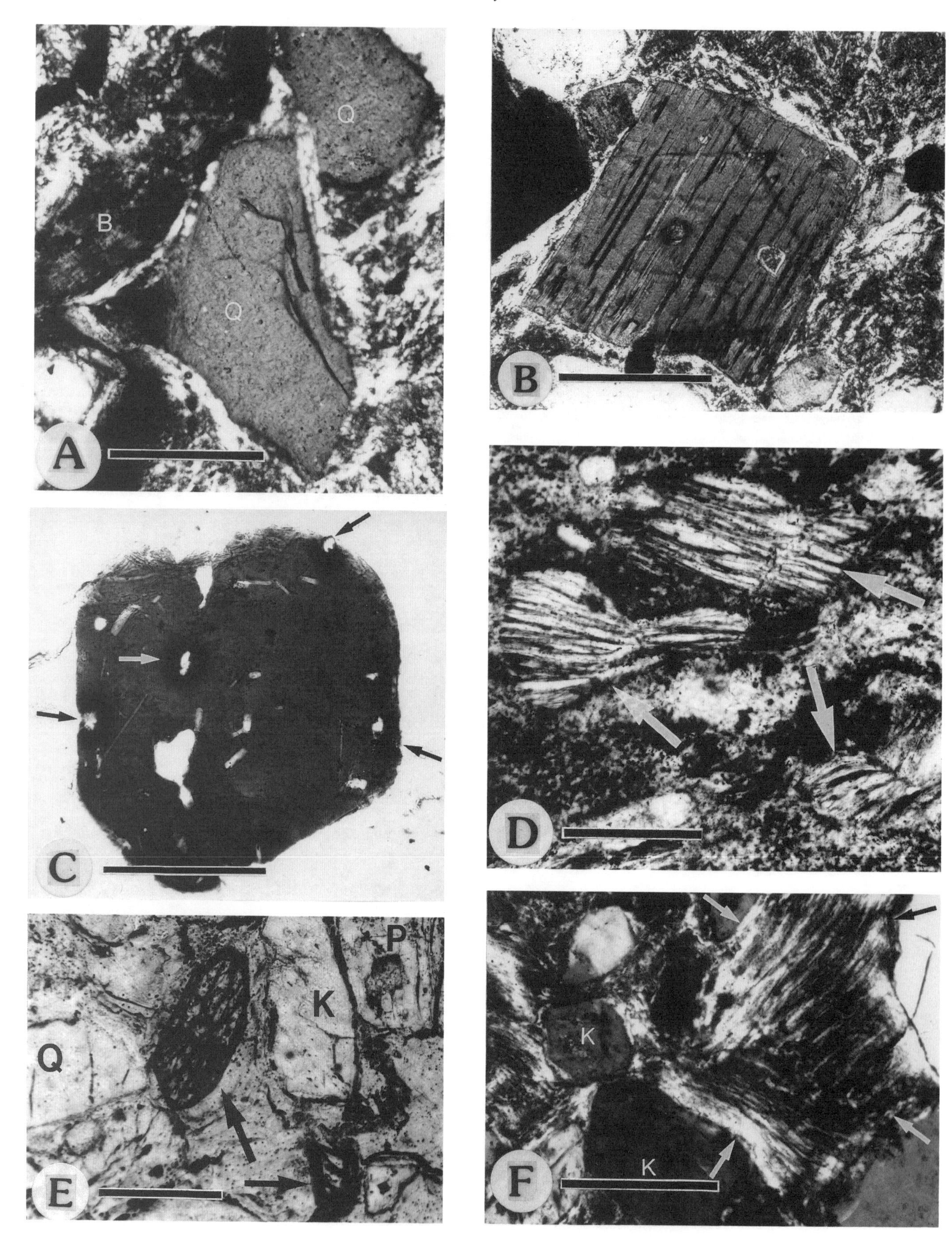
Q
B
Q
A
B
C
D
Q
K
P
E
K
K
F

Figure 10. Biotite, quartz, and other minerals in the Millbrig. The matrix is mixed-layer illite/smectite in all samples. Scale bar is 0.5 mm long. The section numbers are for the Appendix 1 locality register. A, Two quartz phenocrysts (Q). The texture of these grains is typical of quartz phenocrysts in all Millbrig samples. Adjacent grain is biotite (B). From the lowest coarse grained zone in the Millbrig, Carters-Hermitage contact, Shelbyville, Tennessee, section 20 (cf. Fig. 2B). Cross-polarized light. B, Biotite showing the "birch-bark" texture characteristic of micas when viewed in thin section. As is typical of most biotites in the Millbrig, minor alteration to chlorite has occurred along the partings. Diagonal lines are polishing scratches. From the Millbrig, Stones River Formation, Big Ridge, Alabama, section 25. Cross-polarized light. C, Apatite and zircon grains embedded in a large biotite grain. Pleochroic halos occur around the four zircons (arrows). This thin section was cut parallel to bedding, so the euhedral, pseudohexagonal outline of the monoclinic biotite grain is clearly seen. From the Millbrig, Tyrone Limestone core, Marion County, Kentucky, section 41. Plane-polarized light. D, Altered biotites exhibiting a vermicular texture. These grains have a significant chlorite and kaolinite component. From the Millbrig, Bays Formation, Chatham Hill, Virginia, section 10. Plane-polarized light. E, Large euhedral zircon with smaller broken zircon (arrows). Note that the euhedral zircon (center) is nearly as large as the adjacent K-feldspars (K), quartz (Q), and plagioclase (P) grains. From the Millbrig, Tyrone Limestone core, Casey County, Kentucky, section 42. Plane-polarized light. F, A probable relict pumice fragment (arrows) that is now mixed-layer I/S. The space occupied by this grain is significantly larger than the space occupied by the adjacent quartz and K-feldspar (K) grains, suggesting that this is a primary texture. Samples from this particular locality contain an unusual abundance of these grains. From the Millbrig, Carters facies of the Chickamauga Limestone, Tidwell Hollow, Alabama, section 27. Cross-polarized light.

TABLE 3. CHEMICAL ANALYSES OF MILLBRIG BIOTITES*

Sample	BRM-T4	TN 11-1	AL 9-1	AL 14-1
Volumes in wt. %				
SiO_2	35.50	36.09	35.49	36.13
TiO_2	4.07	3.60	4.25	3.05
Al_2O_3	13.94	13.59	13.97	14.28
FeO	25.70	21.56	25.12	17.75
MnO	0.43	0.46	0.36	0.00
MgO	8.18	8.77	8.57	16.14
CaO	0.01	0.00	0.01	0.18
Na_2O	0.36	0.70	0.36	0.72
K_2O	8.90	8.56	8.78	1.38
Total	97.09	93.32	96.91	89.63

*Analyses have not been recalculated to include H_2O as part of the structural formula. Sample localities are given in Appendix 1.

pated that eventually enough biotite grains will be accumulated to obtain a mean value for Deicke samples as well as for the biotite-rich Millbrig samples; the Deicke data could then be added to Figure 11. Unfortunately, so few biotite grains have been observed in thin sections of Deicke samples that at present a meaningful comparison between the Deicke and Millbrig based on microprobe analysis of biotites is not yet possible, and thus the composition of the primary plagioclases (Fig. 8) remains the most certain means of distinguishing the two beds quantitatively when comparing the compositional variations of individual volcanogenic minerals. Certainly the findings of Samson et al. (1988) deserve further study as well because of the potential for apatite and zircon to act as useful discriminators, but it is very important that such studies be based on data from samples collected from an adequate number of localities to ensure a meaningful data set less subject to question (Haynes and Huff, 1990).

Comparison of biotite compositions using the method of Desborough et al. (1973) or of Yen and Goodwin (1976) would be an additional way of directly comparing the Millbrig with the biotite-rich "Big Bentonite" of northern Europe, as part of the ongoing efforts to validate the proposed trans-Iapetus (now trans-Atlantic) correlation of these two beds (Huff et al., 1992).

Accessory minerals. All Deicke and Millbrig thin sections examined contain less than 1% zircon and apatite phenocrysts. Individual crystals typically exhibit euhedral faces, a reflection of their volcanic origin. The zircons are colorless to pale yellow with high relief in plane polarized light. Under cross-polarized light the characteristic high-order interference color is seen. Apatite crystals are also colorless in plane polarized light, but under cross-polarized light they have a low-order interference color. Both the zircons and apatites are unaltered (W. D. Huff, personal communication), as would be expected in an unmetamorphosed sedimentary sequence. They are the smallest crystals observed, with an average length between 0.1 and 0.5 mm. Measurements indicate that both commonly occur in prismatic forms with a length-to-width ratio of about 3:1 up to about 8:1 for zircons, and from 2:1 to 3:1 for apatites. Apatite and zircon grains are commonly embedded in biotite grains. Figure 6B shows a long zircon protruding from a biotite grain, and Figure 10C shows a large biotite grain studded with several apatite and zircon grains. The euhedral cross section of many apatites that have been cut parallel to the *a* and *b* axes is apparent in Figure 10C, and the pleochroic halos around the zircons are visible. Figure 10E shows an exceptionally large euhedral zircon grain that is about as large as the adjacent feldspars; most zircons are considerably smaller.

Secondary calcite and gypsum are also accessory minerals in the Deicke and Millbrig. Calcite is commonly a replacement of primary plagioclase grains (Fig. 7B), and it also occurs as blebs and patches surrounded by clay. It is present in some Deicke samples in abundances up to 15 to 20%, but it rarely makes up more than 5% of Millbrig samples. This is expected, given the tendency for the labradorites in the Deicke to alter more readily than the andesines in the Millbrig, a process that releases abundant Ca once the plagioclase lattice is de-

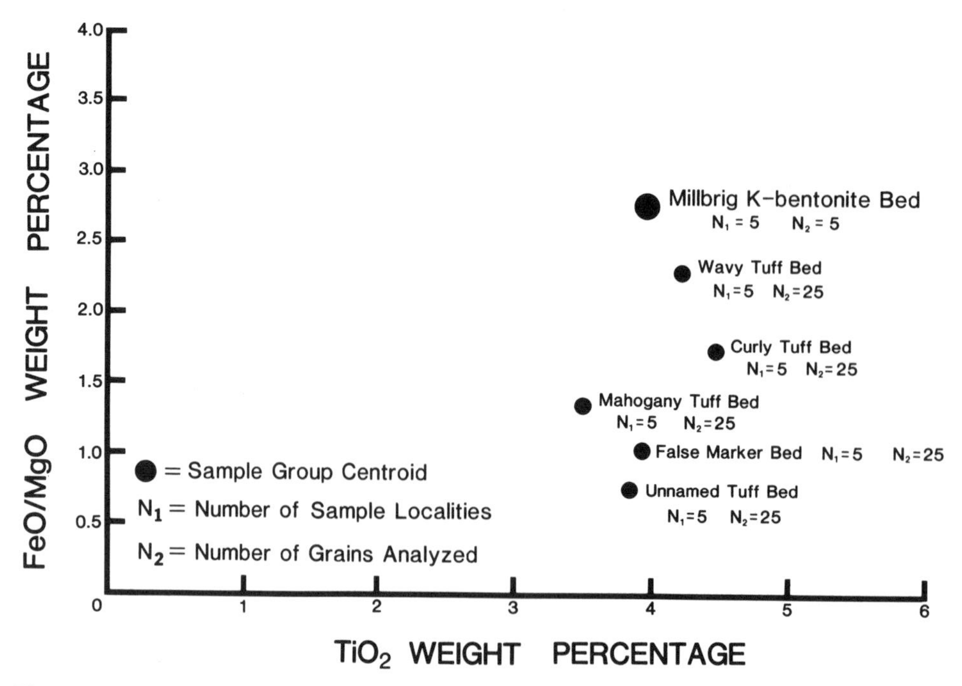

Figure 11. Arithmetic means of FeO/MgO plotted against TiO_2 as determined by microprobe analysis of biotites from seven selected Millbrig samples from five localities in Kentucky, Tennessee, and Alabama. Also plotted are the means calculated for five tuff beds studied by Desborough et al. (1973, Table 1). Following their procedure, two to four measurements were taken on each Millbrig biotite grain analyzed. The use of compositional variation among individual volcanogenic minerals in K-bentonites (apatite, zircon, etc.) as a correlation tool has much potential (Samson et al., 1988), and the above data suggest that the composition of biotites in K-bentonites may also be a useful single-grain correlation tool.

stroyed. Calcite is more common in outcrop samples than cores, reflecting the greater degree of weathering near the surface. It is nearly absent in samples collected from the clastic sections of the eastern Valley and Ridge, but in the carbonate sections farther west calcite is ubiquitous in Deicke samples.

Gypsum was observed in thin sections of one Deicke sample from the Carters Limestone of central Tennessee and one Deicke sample from the Chickamauga Limestone of central Alabama. It occurs as small clusters of bladed, prismatic crystals with a radial habit that were erroneously reported to be zeolites by Haynes et al., (1987). The gypsum crystals have formed in voids and do not appear to be replacing a preexisting mineral.

Clay mineralogy

Mixed-layer illite/smectite. The clay minerals in the Deicke and Millbrig were studied by x-ray diffraction of the <2-μm size fraction. The predominant clay is mixed-layer I/S, which is characteristic of Paleozoic K-bentonites (Huff et al., 1988). Examples of the tracing patterns obtained from x-ray diffraction of oriented, ethylene-glycol–solvated Millbrig samples from the Cincinnati Arch, the carbonate sequences along the western Valley and Ridge, and the clastic sequences along the eastern Valley and Ridge are shown in Figure 12, with supporting data

Figure 12. X-ray diffraction tracings of the < 2-μm size (clay) fraction of selected ethylene glycol–solvated Millbrig samples that show the increase in illite content and ordering from the Cincinnati Arch toward the Valley and Ridge. The section numbers are for the Appendix I locality register. Mixed-layer I/S is the predominant clay mineral. D-spacings for the peaks are shown, and the peak positions in °2Θ are in Table 4. A, Tracings of samples from Kentucky and Virginia. Whereas sample KRI 5-0 (Section 32, Tyrone Limestone) contains about 25% expandable component (smectite) in the I/S, sample VA:WY 1-1 (Section 13, Bays Formation) contains only about 5% smectite. Except for the low-angle "shoulder" which indicates that the 001 peak is really the 001/004* reflection of I/S (Moore and Reynolds, 1989) and thus that an expandable component is present, this tracing is otherwise very similar to a tracing for pure illite. Sample VA 4-1 is from Section 1, Eggleston Formation, and VA:WY 2-1 is from Section 12, Bays Formation, both in the Valley and Ridge. B,. Tracings of samples from Tennessee, Georgia, and Alabama. Sample TN 7-3 (Section 20, Carters Limestone) is compositionally very similar to sample KRI 5-0 in A, as is expected because both are from Cincinnati Arch sections. Appreciable kaolinite (k) is present along with the I/S in sample AL:CH 1-4 (Section 28, Colvin Mountain Sandstone). Sample GA:WL 3-2 is from Section 22, Carters Limestone, in the western Valley and Ridge of Georgia.

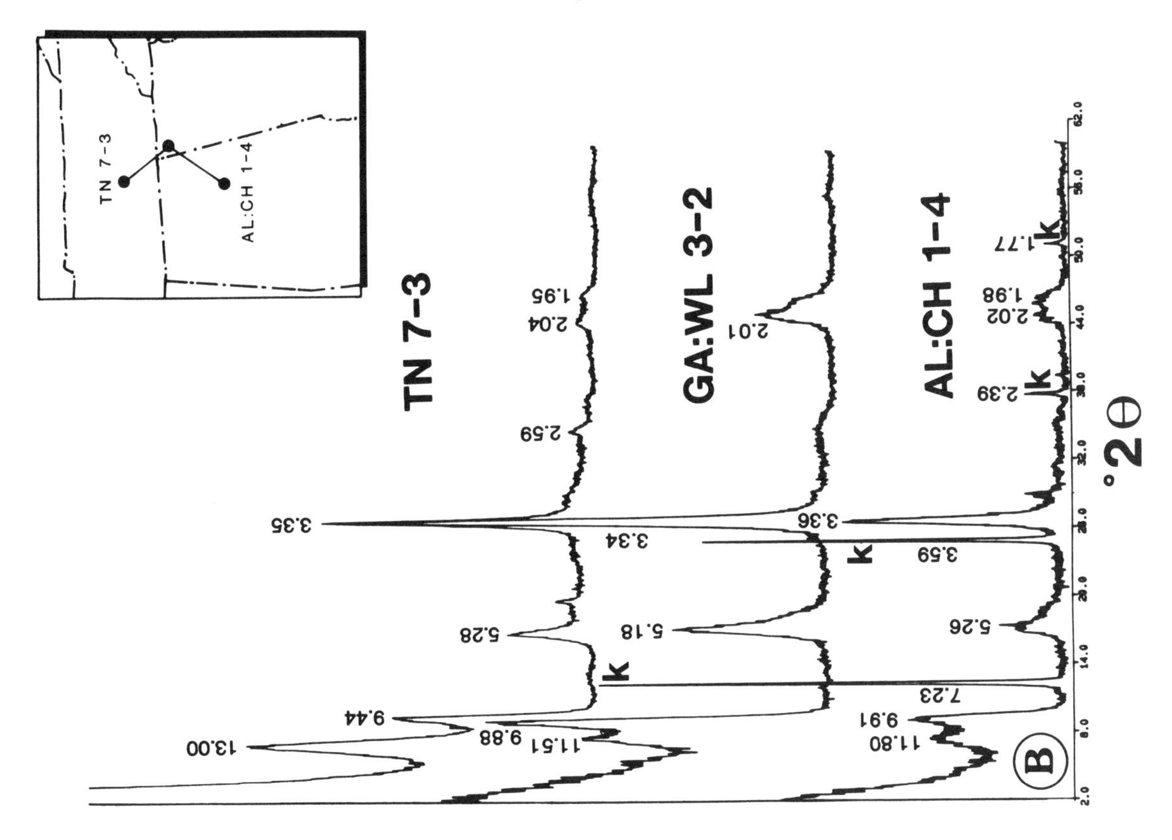
TN 7-3
AL:CH 1-4
TN 7-3
GA:WL 3-2
AL:CH 1-4
B
°2Θ

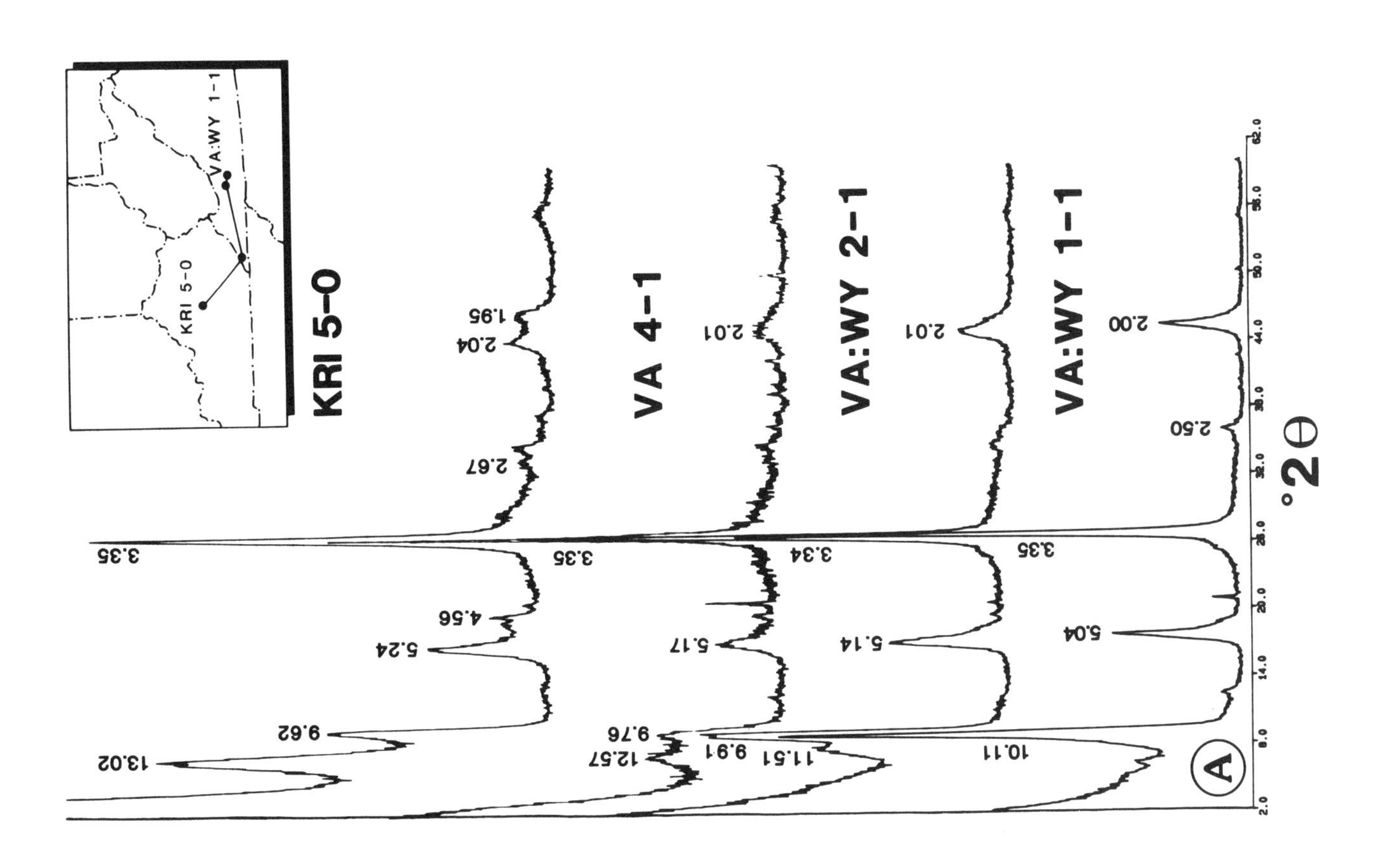
KRI 5-0
VA:WY 1-1
KRI 5-0
VA 4-1
VA:WY 2-1
VA:WY 1-1
A
°2Θ

presented in Table 4. Because the clay mineralogy of the Rocklandian K-bentonites along the Cincinnati Arch has been investigated in detail (Huff, 1963; Lounsbury and Melhorn, 1964; Günal, 1979; Huff and Türkmenoğlu, 1981), all but five x-ray diffraction analyses were of samples from the Valley and Ridge, where study of the clays has been much less detailed (Reade, 1959; Coker, 1962; Hergenroder, 1966; Lounsbury and Melhorn, 1964, 1971; Manley and Martin, 1972; Elliott and Aronson, 1987).

In thin section, the matrix of the K-bentonites, which is the I/S, is highly birefringent and microcrystalline. In samples cut parallel to bedding the matrix has a grid-like texture, as shown in Figure 6A, where it surrounds the plagioclase grain. In thin sections cut normal to bedding, it is evident that the highly birefringent I/S matrix has been squeezed around feldspars, opaques, and the other relatively large nonplaty grains.

Using the graphic method of Watanabe (1981) as modified by Środoń and Eberl (1984), the percentage of smectite and the ordering of the I/S was determined graphically from the x-ray diffraction data in Table 4 (Fig. 13). The Watanabe method is a plot of two parameters derived directly from the diffraction data, the differential measurements $\Delta 2\theta_1$ and $\Delta 2\theta_2$ (Środoń and Eberl, 1984). The use of differential measurements can minimize the effect that variations in the interlayer ethylene glycol solvation complex have on the interpretation of the diffraction data because reflections that are affected by solvation are displaced in the same direction, meaning that the absolute distance between them remains nearly unchanged. Also, results based on a plot of differential measurements are ". . . relatively insensitive to goniometer zero-alignment problems and specimen displacement errors." (Moore and Reynolds, 1989, p. 251). Finally, problems that are introduced by the presence of small amounts of quartz are greatly reduced when differential measurements are plotted. For these reasons, the graphic method of Watanabe (1981) was used.

The ordering of an I/S sample is an indication of the illite:smectite ratio. R0 (random) ordered samples contain 10 to 50% illite, R1 (short-range) ordered samples contain 50 to 85% illite, and R3 (long-range) ordered samples contain greater than 85% illite (Reynolds and Hower, 1970; Reynolds, 1980). The ordering in the Deicke and Millbrig changes gradually toward the southeast, from R1 along the Cincinnati Arch to R1/R3 at a few of the westernmost localities in the Valley and Ridge, and then to R3 in most of the Valley and Ridge. In Figure 12A, the peak pattern obtained for sample KRI 5-0 following ethylene glycol solvation shows a well-defined 001 smectite peak (d-spacing 13.02) and 001 illite peak (d-spacing 9.62). Graphic analysis (Fig. 13) shows that this and other samples from the Tyrone and Carters Limestones along the Cincinnati Arch contain up to 30% smectite. By contrast, the peak pattern for the easternmost sample in Figure 12A, VA:WY 1-1, from the Bays Formation in the eastern Valley and Ridge of Virginia, shows a well-developed 001 illite peak (d-spacing 10.11). That pattern is very close to the diffraction pattern for pure illite except that both the 001 and 002 peaks are slightly asymmetric, with a very weak shoulder on the low-angle side of the 001 peak and on the high-angle side of the 002 peak, features that identify this as a diffraction pattern of highly illitic I/S, not pure illite (Środoń and Eberl, 1984; Moore and Reynolds, 1989). Graphic analysis shows that this sample in fact does contain about 3% smectite.

The same regional trend, an overall increase in ordering from northwest to southeast, is evident in sections farther south (Fig. 12B). Note that the I/S in sample AL:CH 1-4, from the Colvin Mountain Sandstone of Alabama, contains a higher percent of expandable layers than does the I/S in the samples from the Bays Formation, the Colvin Mountain equivalent, in Virginia (samples VA:WY 2-1 and VA:WY 1-1 in Fig. 12A). Also, the maximum ordering in samples from Alabama and Georgia is less than the maximum in the equivalent strike belts in Virginia, and overall the most illitic K-bentonites studied are in west-central Virginia.

Other clays. In the Colvin Mountain Sandstone, both the Deicke and Millbrig contain abundant, well-crystallized kaolinite, the diffraction peaks of which are clearly evident in the x-ray diffraction pattern for sample AL:CH 1-4 (Fig. 12B). Kaolinite is also present in minor amounts in Millbrig samples from the Bays Formation in Virginia and Tennessee. It was identified on some x-ray diffraction tracings but could not be distinguished in the matrix in thin section. Instead, the kaolinite appears to be associated with many of the biotites in the Bays. Those grains, which are partly chloritized, have also been partly altered to kaolinite, and perhaps vermiculite, along 001 cleavage planes. This alteration has produced an expanded, vermiform texture, and a more random alignment of these grains instead of the usual alignment parallel to bedding (Fig. 10D). Because those grains still contain some biotite layers, they may be a biotite-kaolinite aggregate, or a mixed-layer biotite/vermiculite, which is known as hydrobiotite. Biotites that have altered to biotite-kaolinite aggregates are present in Silurian K-bentonites of Gotland (Snäll, 1977). Those Silurian biotites have a vermiform texture and are oriented at random relative to bedding, whereas separate, unaltered biotite grains are oriented parallel to bedding. Photomicrographs in Snäll (1977) show a texture that is remarkably similar to what is seen in the Millbrig samples from the Bays. Additional study of these grains is planned.

An unusual grain observed in one sample of the Millbrig is shown in Figure 10F. This is a single, optically continuous "grain" of I/S having a sweeping extinction pattern. It occupies a larger space than adjacent feldspar and quartz phenocrysts, and thus it is likely that that space most likely was originally occupied by a large pumice fragment deposited with the smaller but hydraulically equivalent quartz and feldspar phenocrysts. Subsequent alteration and devitrification of the pumice fragment has produced this single grain of clay, and a

TABLE 4. I/S PEAK POSITIONS, PERCENTAGE OF ILLITE, AND ORDERING BASED ON GRAPHIC ANALYSIS OF X-RAY DIFFRACTION TRACINGS*

Sample	Peak Positions (°2θCuKα)						Average Illite (%)	Ordering
				DEICKE				
Cincinnati Arch								
BB 5-5-1	6.898	9.173	16.869	26.648	44.639	46.570	77	R1
Valley and Ridge: Carbonate sections								
AL 7	7.087	8.926	17.148	26.705	45.174	....	87	R1/R3
GA:DD 2-1	7.433	9.030	17.198	26.711	45.055	....	87	R1/R3
TN:CL 1-1	....	8.769	17.386	26.689	45.285	....	92	R3
VA:SC 1-1	....	8.666	17.480	26.514	44.998	....	92	R3
VA:RL 2-1	6.990	9.036	17.009	26.573	44.896	....	83	R1/R3
VA:TZ 1-7	....	8.744	17.261	26.549	45.059	....	90	R3
VA:BL 1-1	....	8.768	17.426	26.634	45.202	....	92	R3
VA:GI 4-0	....	8.815	17.404	26.654	45.245	....	93	R3
Valley and Ridge: Colvin Mountain Sandstone sections								
AL:CH 1-2	7.445	9.102	17.100	26.654	44.916	....	85	R1/R3
AL:ET 1-2	....	8.771	17.694	26.700	45.454	....	98	R3
GA:FL 1-1	....	8.775	17.535	26.662	45.425	....	95	R3
				MILLBRIG				
Cincinnati Arch								
BB 5-2-3	6.910	8.936	16.983	26.671	44.672	46.227	80	R1
KRI 5-0	6.784	9.183	16.897	26.606	44.330	46.595	76	R1
TN 7-3	6.971	9.362	16.791	26.572	44.394	46.576	75	R1
TN 7-4	6.816	9.080	16.814	26.505	44.380	46.971	79	R1
Valley and Ridge: Carbonate sections								
GA:DD 2-2	....	9.214	16.833	26.655	44.686	46.608	76	R1
GA:WL 3-2	7.541	8.829	17.172	26.646	44.971	....	87	R3
TN:CL 1-2	7.224	8.817	17.201	26.614	45.223	....	88	R3
VA 4-1	7.027	9.050	17.138	26.597	44.966	....	86	R1/R3
VA:LE 1-2	....	8.543	17.284	26.673	44.663	....	87	R3
VA:SC 1-2	....	8.785	17.287	26.662	45.115	....	90	R3
VA:GI 1-1	....	8.741	17.450	26.551	45.198	....	94	R3
Valley and Ridge: Colvin Mountain Sandstone and Bays Formation sections								
AL:CH 1-4	7.400	9.150	16.849	26.495	44.731	46.091	80	R1/R3
VA:SM 1-4	7.536	9.022	17.212	26.610	44.897	....	88	R3
VA:WY 2-1	7.646	8.920	17.224	26.646	45.064	....	88	R3
VA:WY 1-1	....	8.740	17.588	26.607	45.267	....	97	R3
VA:RO 1-3	7.600	9.032	17.118	26.662	44.810	....	85	R3
VA:BT 2-1	7.710	8.715	17.131	26.663	45.289	....	89	R3
				K-BENTONITE BED V-7				
Valley and Ridge: Carbonate section								
VA:TZ 2-3	....	8.702	17.160	26.531	44.717	....	87	R3
Valley and Ridge: Bays Formation sections								
VA:RO 1-2	7.671	8.960	17.176	26.615	44.883	....	86	R3
VA:BT 2-2	7.600	8.849	17.536	26.662	45.420	....	94	R3
				OTHER K-BENTONITES				
Valley and Ridge: Carbonate sections								
AL:ET 2-4	7.177	8.922	16.948	26.646	44.930	....	81	R1/R3
AL:ET 2-5	7.003	8.797	17.287	26.653	45.293	....	91	R1/R3
GA:WL 3-3	7.415	8.769	17.210	26.641	45.173	....	88	R1/R3

*The average illite percentage and the ordering were calculated from x-ray diffraction tracings using the technique summarized by Srodon and Eberl (1984). R1/R3 ordering of Srodon, used herein, is the same as R2 ordering of Watanabe (1981). Sample localities are given in Appendix 1.

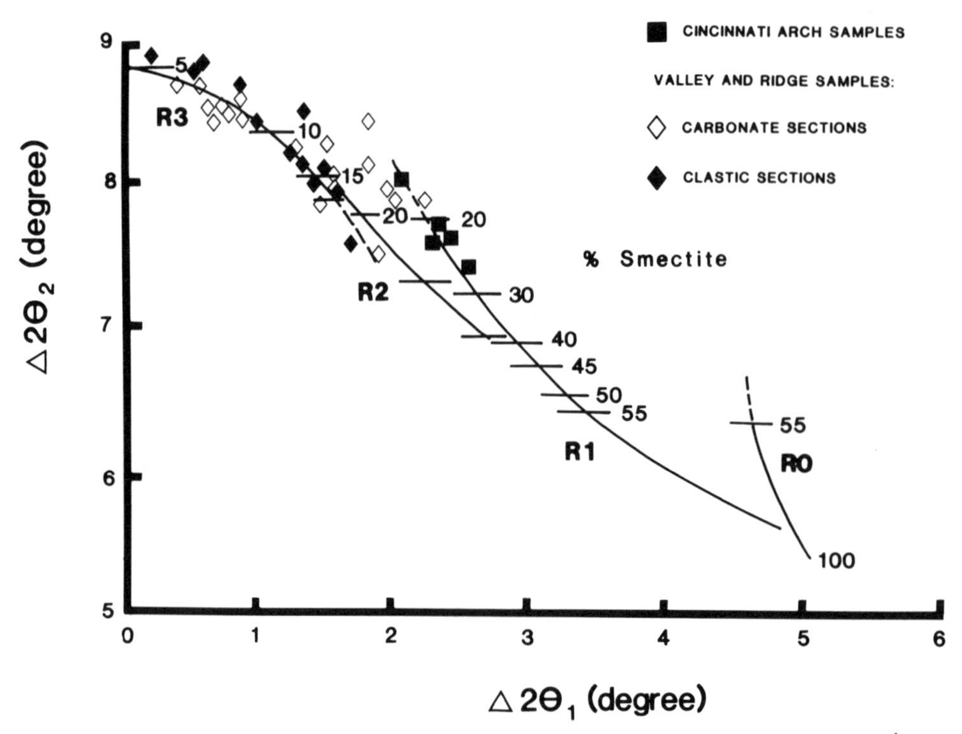

Figure 13. Results using the graphic analysis method of Watanabe (1981), as modified by Środoń and Eberl (1984), to determine the composition and ordering of mixed-layer I/S in the Deicke and Millbrig. R2 ordering is considered R1/R3 ordering by Środoń and Eberl (1984), whose terminology is used herein except on this graph. The derivation of $\Delta 2\Theta_1$ and $\Delta 2\Theta_2$ is discussed by Watanabe (1981) and Środoń and Eberl (1984), and the advantages of using differential measurements for this type of determination are discussed by Moore and Reynolds (1989).

texture that was not destroyed by postdepositional compaction of the bed.

Color. At a given exposure, the Deicke and Millbrig are usually not the same color. In samples from outcrops and cores along the Cincinnati Arch, the Deicke is various shades of yellow to yellow-green to green to blue-green, whereas the Millbrig is more greenish gray to gray. These are also the colors of samples collected in the limestones of northeast Alabama, northwest Georgia, and southeast Tennessee, and in the Powell Valley of Tennessee and southwest Virginia. In Virginia and Tennessee where the Millbrig occurs in the Eggleston Formation and the Deicke occurs in or just above the Moccasin Formation, both argillaceous carbonate units, Millbrig samples tend to be more brownish yellow, whereas Deicke samples are red to reddish brown with yellow lenses. Samples of both K-bentonite beds from the Colvin Mountain Sandstone in Alabama and Georgia are a pale grayish or greenish white. Millbrig samples from the Bays Formation in Virginia and Tennessee are brownish red to brownish yellow; the Deicke does not occur in the Bays. In addition, samples of the Deicke and Millbrig from the Upper Mississippi Valley are brown to orange in color (Willman and Kolata, 1978; Kolata et al., 1986).

These variations are caused by differences in the color of the clay matrix, as indicated by the surprisingly diverse array of colors in the clay mineral mounts that were prepared for x-ray diffraction. In almost all clay-rich sediments such as shales and other mudrocks, color is dictated primarily by variations in the amounts of ferrous and ferric iron and organic material (Potter et al., 1980). The color differences of the K-bentonites—which, like shales, are mostly fine-grained, clay-rich rocks—are, as with shales, caused by differences in the oxidation state of the iron oxides that are adsorbed onto the clays.

PETROLOGY AND PETROGENESIS

Introduction

In hand samples and in thin sections, the presence of abundant biotite and quartz immediately distinguishes the Millbrig from the Deicke, which is itself recognizable by the presence of abundant Fe-Ti minerals and the near total lack of quartz and biotite. Based on these mineralogic differences and on the differences in primary plagioclase compositions (Fig. 8), some inferences about the original composition of the volcanic ash that has since altered to the Deicke and Millbrig K-bentonites can now be made. Although postdepositional changes such as the alteration of original glass and the ubiquitous alteration of feldspars mask many of the clues that aid in a petrologic description of younger or less altered volcanic rocks, those changes have not destroyed all the clues that are helpful in constraining the composition of the original ashes.

Nature of the original ashes

Phenocryst mineralogy indicates that the Deicke and Millbrig K-bentonites are derived from ashes of acid to intermediate composition. The quartz and andesine in the Millbrig indicate derivation from a felsic ash, and the abundant biotite narrows the choice further; the original ash was most likely a rhyodacite. Although rhyodacites and other felsic ashes commonly contain both plagioclase and K-feldspar as primary phenocrysts, it is likely that the near total lack in the Millbrig (and in the Deicke) of a primary K-feldspar such as sanidine does *not* reflect the composition of the original magma. The presence of biotite and quartz indicates that the magma contained abundant silica, aluminum, and potassium. Presumably sanidine phenocrysts, which are now very rare in the K-bentonites, were originally present but have since been removed by postdepositional processes.

Although biotite in volcanic rocks is commonly highly altered and resorbed, the magma from which the Millbrig ash was derived was more hydrous than many magmas, as indicated by the presence of relatively unaltered biotites in many samples. It is evident that, prior to eruption, the magmatic conditions favored the formation of biotite over other mafic ferromagnesian silicate minerals. Certainly the hydrous nature of the magma would have favored biotite over pyroxene, and evidently the prevailing P, T, and X conditions favored the formation of biotite over an amphibole such as hornblende. No trace of amphiboles, pyroxenes, or other higher temperature mafic phenocrysts was observed in samples of the Millbrig, either as resorbed grains, highly altered cores, or otherwise.

The presence of labradorite and ilmenite in the Deicke emphasizes the evidently high content of CaO and TiO_2 in the original ash, and this, together with the almost total lack of quartz and biotite, indicates that the original Deicke ash was of a more intermediate composition than the Millbrig ash. The Deicke is most likely derived from a dacitic or perhaps latitic ash.

Andesites, basalts, and other intermediate to mafic extrusive igneous rocks commonly contain ferromagnesian minerals such as olivine, amphibole, pyroxene, or biotite in varying abundances. Essentially all of the magnesium in the Deicke, however, is now in the illite/smectite, and the most abundant ferroan minerals in the Deicke are not ferromagnesian minerals but the ferrotitanian mineral ilmenite, and the TiO_2 polymorphs that are pseudomorphic after the ilmenite. Biotite is present but very rare in the Deicke, and there are no resorbed grains or highly altered cores of olivine, amphibole, or pyroxene, although some of the TiO_2 minerals appear to be pseudomorphic after pyroxene rather than ilmenite (Fig. 9C).

The Millbrig likely contained primary K-feldspar phenocrysts (now gone) in addition to the quartz, biotite, and andesine phenocrysts still present, and the Deicke likely contained primary K-feldspar and rare pyroxene phenocrysts (now gone), in addition to the labradorite and ilmenite phenocrysts still present. A comparison of these inferred phenocryst assemblages with the assemblages of some well-studied Cenozoic volcanic ashes (Heiken, 1972; Scarfe et al., 1982; Fisher and Schminke, 1984; Self et al., 1984; Rose and Chesner, 1987; Best et al., 1989) supports the conclusion that the Millbrig is altered rhyodacitic ash and the Deicke is altered dacitic or latitic ash.

The composition of the original glass remains speculative for both the Deicke and Millbrig, although trace element distribution can give additional insight into their composition. Based on a plot of Zr/TiO_2–Nb/Y ratios obtained from analysis of the clay fraction only, Huff and Türkmenoğlu (1981) suggested that some Ordovician K-bentonites from along the Cincinnati Arch were derived from a trachyandesitic ash, but others were derived from rhyodacitic/dacitic and rhyolitic ashes. Their data set included samples from the beds that are now recognized as the Deicke and the Millbrig. Consideration of phenocrysts suggests that the Millbrig is not derived from a trachyandesitic ash, but such an ash cannot be ruled out as a precursor of the Deicke given the near absence of quartz and the probability that sanidine phenocrysts were originally present.

Volume and areal extent of the original ashes

Individual exposures of the K-bentonite beds contain important clues about the relative duration and volume of the eruptive events that produced the original ashes. Figures 14 through 19 show details of the internal stratigraphy of the Deicke and Millbrig at selected outcrops. At the Big Ridge, Alabama, section (Fig. 15), the Deicke overlies a prominent chert layer in the lime mudstones of the Stones River Group. This lower contact is very sharp. Above this is a 1- to 2-cm-thick layer of phenocryst-poor, plastic clay, followed by a zone of coarse-grained tuffaceous material that grades upward into successively finer grained, more bentonitic material. The upper contact of the bed is much less sharp than the lower contact, presumably because the upper few centimeters of ash were reworked into the overlying sediments. The texture overall is thus that of a normally graded bed except for the basal layer of plastic clay. This vertical sequence characterizes the Deicke in exposures along the Cincinnati Arch and in the western Valley and Ridge from the Powell Valley in Virginia to the Big Wills Valley in Alabama, and it suggests that the Deicke ash was erupted during a single, extremely large eruption.

In the Deicke at the Gate City, Virginia, section (Fig. 16) the phenocrysts are much smaller than in exposures farther south, but the thin plastic clay at the base seen at Big Ridge is still present. Where the Deicke occurs in the Moccasin redbeds of southwest Virginia and northeast Tennessee, as at

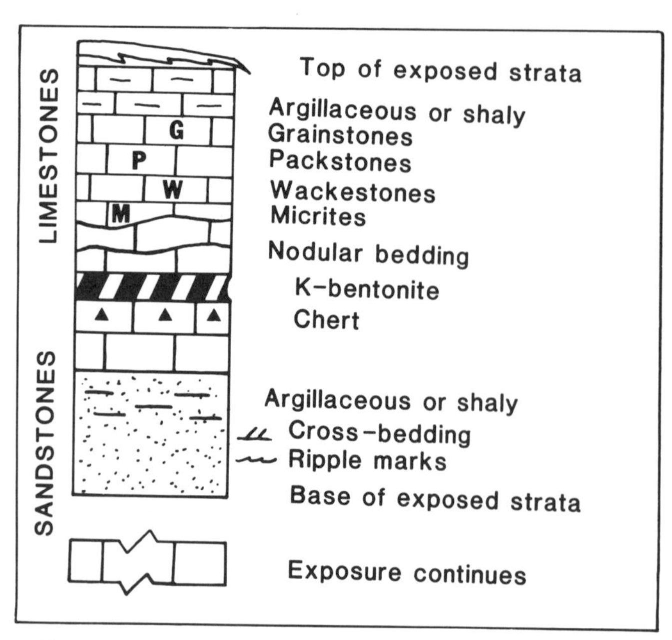

Figure 14. Legend for Figures 15 through 19, and 25 through 33.

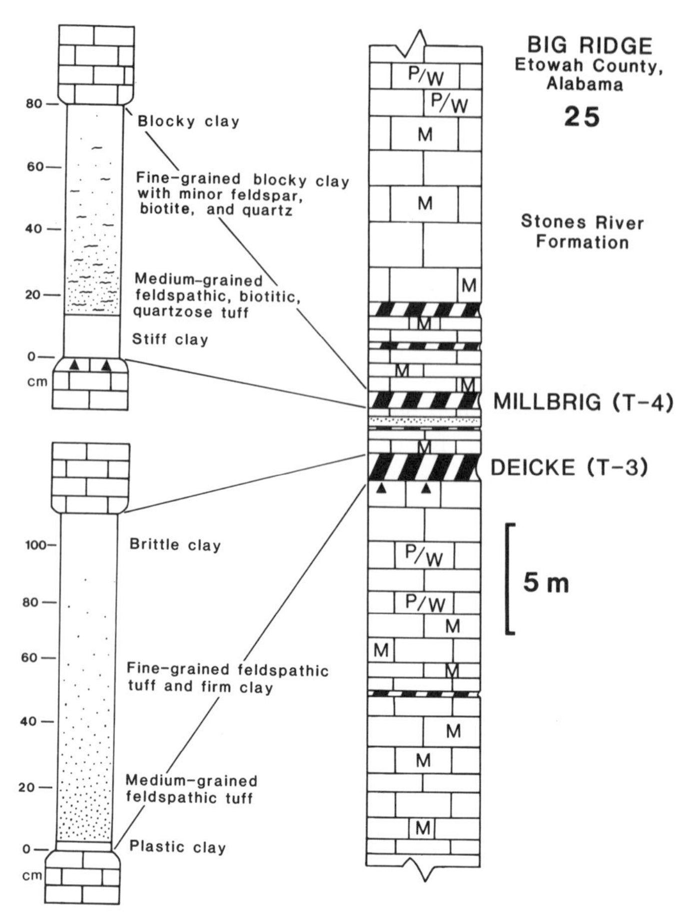

Figure 15. Details of the measured section along I-59 in the cut through Big Ridge near Duck Springs, Alabama. The vertical variations shown are typical of both the Deicke and Millbrig throughout the Cincinnati Arch and the carbonate sequences of the western Valley and Ridge from Alabama to Virginia. The coarse zone in the Deicke is consistently the basal 10 to 15 cm of the bed, whereas the coarse biotite-rich zone in the Millbrig is consistently in the interval 8 to 20 cm above the base of the bed, distinct textural differences that were succinctly and accurately described 50 yr ago by Bay and Munyan (1940) and Fox and Grant (1944) but too often ignored since. The thin sandstone between the Deicke and Millbrig is a subrounded to rounded calcite-cemented mature quartz arenite that is the distal edge of the Colvin Mountain Sandstone shown in Figure 18 (Chowns and McKinney, 1980).

Gate City, the normal grading of the bed is less noticeable than in sections where it is in the gray or greenish gray limestones of the Carters, Tyrone, Chickamauga, or Eggleston Formations. North of Gate City, the thin plastic clay at the base is usually absent, but the Deicke can be readily identified because of its stratigraphic association with the Moccasin redbeds and its own red color, which is in sharp contrast to the Millbrig in a given exposure (Haynes, 1992).

Whereas the Deicke is a simple, normally graded bed, the internal stratigraphy of the Millbrig is more complex. At Big Ridge (Fig. 15) and in the carbonate sequences throughout the region, the Millbrig also typically overlies a prominent chert layer in the underlying lime mudstones; this lower contact is very sharp. At most exposures this contact is followed by 8 to 12 cm of fine-grained clay, which is the lowest discrete zone in the Millbrig. Above this zone is 15 to 20 cm of tuffaceous material. Here are found the coarsest phenocrysts in the Millbrig, and it is this coarse-grained, biotite-rich zone that has been noted in many previous studies (Bates, 1939, Bay and Munyan, 1940; Fox and Grant, 1944; Huffman, 1945; Miller and Fuller, 1954). This tuffaceous layer grades upward into successively finer grained and thus more bentonitic material. The diagnostic phenocrysts of the Millbrig, especially biotite and quartz, are found throughout this upper zone, but their size and abundance decreases steadily upward. The upper contact of the bed is, like the Deicke, less distinct than the lower contact because the ash was reworked into the overlying calcareous or argillaceous sediments.

This internal vertical stratigraphy of several distinct layers of altered ash suggests that the Millbrig represents, in many exposures, either the cumulative result of an extended period of eruptive activity, or the cumulative effects of an evolving magma chamber during a single massive eruptive event. This problem is currently under study elsewhere (W. D. Huff, personal communication).

Where the Millbrig occurs in the clastic strata of the Bays Formation in the eastern Valley and Ridge province of Virginia and Tennessee, the sequence at most exposures is very similar to that observed in the carbonate sequences as described above, and the Crockett Cove, Virginia, section is representative (Fig. 17). There, a layer of coarse phenocrysts occurs above the fine-grained zone at the base, but even in the

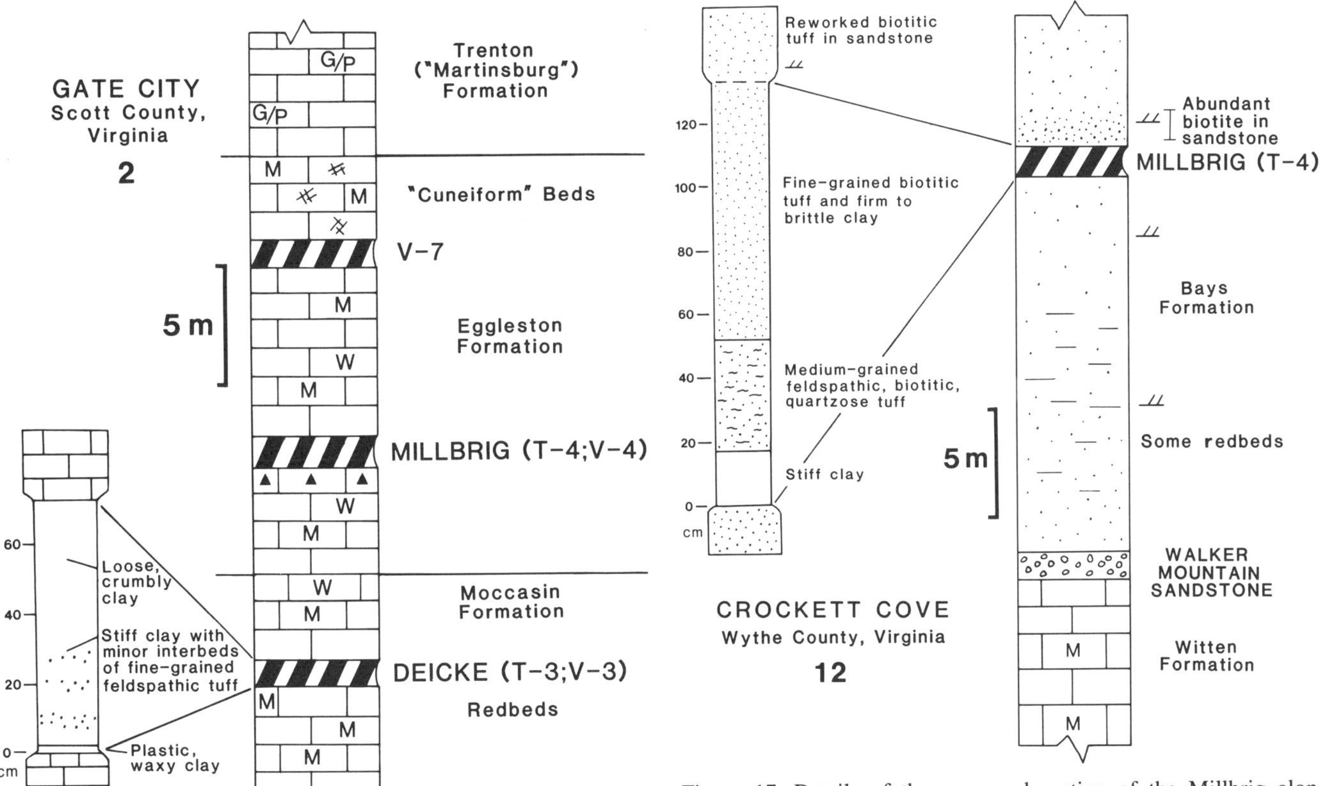

Figure 16. Details of the measured section of the Deicke on Clinch Mountain in Gate City, Virginia. The 1- to 3-cm-thick layer of plastic clay at the very base of the bed is one of the diagnostic textural features of the Deicke in much of the southeastern United States. At the Thorn Hill and Eidson sections on Clinch Mountain south of Gate City, this clay is overlain by a coarse-grained zone several centimeters thick, which contains numerous feldspars and Fe-Ti minerals, but this coarse-grained zone is indistinct at Gate City. Farther north in the Moccasin Formation sections both the basal plastic clay and the coarse-grained zone are absent, but the Deicke can still be recognized because of its distinct red color, its occurrence at or near the top of the Moccasin redbeds, its thickness, and its proximity to the Moccasin-Eggleston contact throughout the region (cf. Fig. 28).

Figure 17. Details of the measured section of the Millbrig along westbound I-77 in the cut through Cove Mountain near Crockett Cove, Wytheville, Virginia. In exposures of the Moccasin Formation to the west the Deicke occurs just above the Walker Mountain Sandstone Member, but here in the Bays Formation where the Walker Mountain is the basal unit of the Bays (Haynes, 1992) the Deicke is not present. This is one of the thickest exposures of the Millbrig in the southern Appalachians.

upper parts of the bed there is still a relative abundance of coarse phenocrysts. The presence of abundant coarse-grained biotite in the sandstones immediately overlying the Millbrig at Crockett Cove, particularly along bedding planes, is evidence that some ash was reworked, and that reworking is thus the best explanation for the poorly defined upper contact of both the Deicke and the Millbrig in nearly all exposures studied.

The exposures in the Colvin Mountain Sandstone of Georgia and Alabama are much different from those in the Bays Formation farther north. At the Alexander Gap section (Fig. 18), both the Deicke and Millbrig are nearly homogenous beds of clay, with no observable vertical grading. All phenocrysts in the Millbrig, including the quartz grains, have evidently been completely altered by postdepositional diagenetic processes. In the Deicke, the only identifiable phenocrysts are the abundant ilmenite grains, although throughout the bed there are numerous small patches or blebs of blue-green clay aligned parallel to bedding that contrast sharply with the surrounding pale gray to white clay. In thin section, those green specks are rectangular and thus may be the relict outlines of former pyroxene grains.

At a quarry near Shelbyville, Tennessee, in the Central Basin, the Millbrig includes the usual clay-rich layer above the base, followed by the tuffaceous biotite-rich layer. Above this, however, are two thin, but very distinct, tuffaceous layers that are normally graded (Fig. 19). Those two layers may represent additional ashfall deposits. They are very thin compared with the lower and much thicker biotite-rich layer and the lowermost clay-rich layer, but their thinness suggests that they would likely have been obliterated by any significant reworking of the ash. The surrounding micrites of the Carters Limestone were deposited in a restricted peritidal environment, and reworking of the ash by either physical or biologic processes was evidently very minor at this exposure. The Shelbyville section provides the clearest evidence that, unlike the Deicke, the

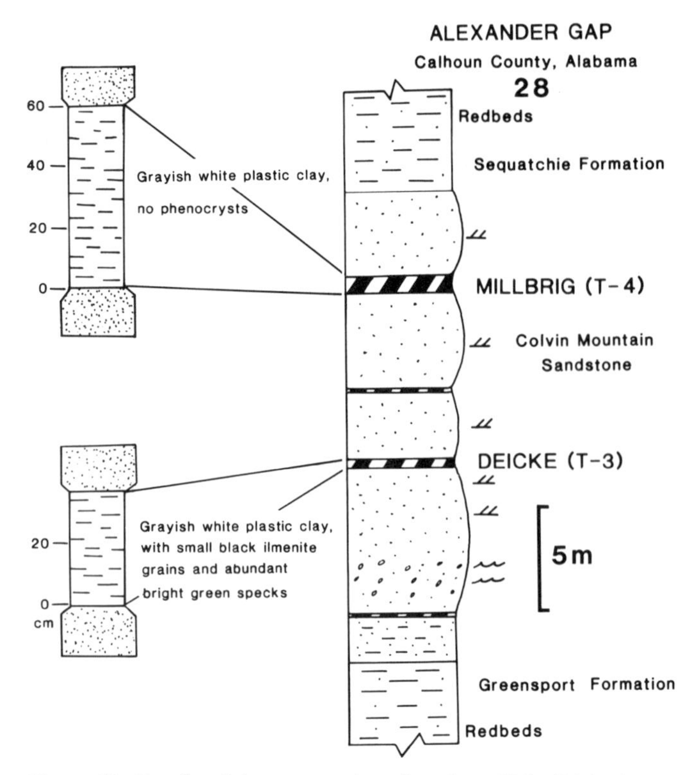

Figure 18. Details of the measured section along U.S. 431 in the cut through Colvin Mountain at Alexander Gap, Alabama. The Deicke and Millbrig occur in the mature quartz arenites of the Colvin Mountain Sandstone. Both beds have been kaolinized to some extent, and overall they are extremely altered. Neither contains any feldspar phenocrysts, biotite and even quartz are absent from the Millbrig, and the Deicke is recognizable only by the presence of small but ubiquitous ilmenite grains (cf. Fig. 9D).

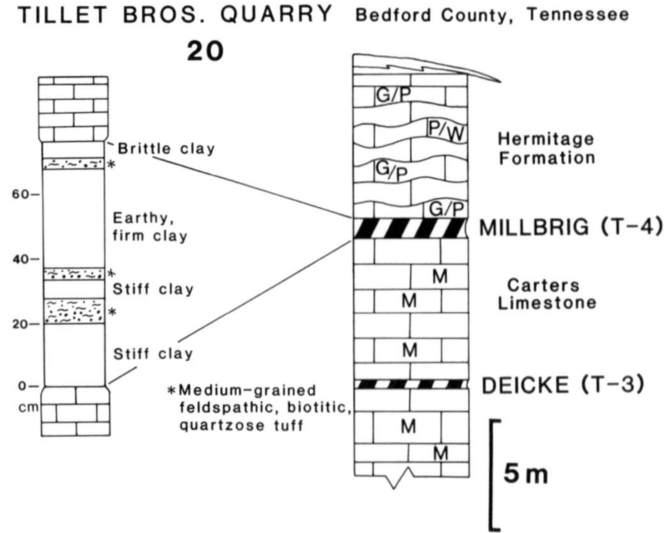

Figure 19. Details of the measured section of the Millbrig in the Tillet Brothers Quarry near Shelbyville, Tennessee. This Millbrig exposure is the only one known to contain the discrete thin, graded, coarse-grained layers above the thicker coarse-grained layer, which is in its normal stratigraphic position several centimeters above the base of the bed. The internal stratigraphy of the Big Bentonite at Kinnekulle, Sweden (Bystrom, 1956; Kunk et al., 1984; Huff et al., 1992), is remarkably similar to the stratigraphy of the Millbrig at this quarry in south-central Tennessee.

Millbrig is composed of more than one ashfall; it further indicates that, at least along the Cincinnati Arch, four separate ashfalls may have accumulated, beginning with a shard-rich ash and followed by at least three crystal-rich ashes. In the Valley and Ridge, where stratigraphic thicknesses suggest that sedimentation rates were greater than along the Cincinnati Arch, those ashfalls may have formed completely separate K-bentonites; the exposure at Big Ridge contains several thinner K-bentonites upsection from the Millbrig (Fig. 15), and biotite is abundant in at least two of those beds. They could conceivably be the same two thin beds that occur in the upper Millbrig at Shelbyville, as those two graded beds were not observed in any other exposures of the Millbrig. It is also very possible that the multiple beds seen at Shelbyville simply represent pulses of ash from a single huge eruptive event lasting several days.

The Deicke and Millbrig ashes were deposited over a large area, estimated to be at least 600,000 km^2 (Huff and Kolata, 1990). The two beds can now be recognized from Birmingham, Alabama, to Daleville, Virginia (north of Roanoke), a linear distance of about 800 km. From Daleville they can be traced west to the Cincinnati Arch and beyond, to Cape Girardeau, Missouri, a linear distance of about 850 km. From there they can be traced north to St. Paul, Minnesota, a linear distance of about 900 km (Kolata et al., 1986; Huff and Kolata, 1990). Thus, each of these K-bentonites is comparable in volume and areal extent to some of the largest ashfalls known, both historically and from the rock record. Fresh ash from the Tambora eruption of 1815, one of the largest eruptions during the last 10,000 yrs, was reported 1,000 km from the vent in accumulations of 1 cm, and at least 175 km^3 of pyroclastic material was erupted in a 24-hr period (Self et al., 1984). By comparison, ash from the Toba eruption (75 Ka) has been found 3,100 km distant, and a minimum of 800 km^3 of ash alone is thought to have been erupted (Rose and Chesner, 1987).

A conservative estimate for the present volume of the Deicke and the Millbrig K-bentonites in North America is 330 km^3 each (Huff and Kolata, 1990). This number is substantially larger for the Millbrig when the volume represented by the "Big Bentonite" of northern Europe is accounted for (Huff et al., 1992). And, when compaction of the precursor ash is considered, the original volume was almost 3.9 times this amount (Dokken, 1987, whose calculated compaction ratio of 3.87 to 1 based on ellipse measurements of burrows was erroneously cited by Huff and Kolata, 1990, as 3.4 to 1). Thus, the two separate eruptive events that produced the Deicke and Millbrig ashes generated a minimum of 1,270 km^3

of ash, so the amount of ash from *each* of those events was greater than that from the Toba eruption. This estimated volume of erupted ash is far greater than that given in the only other known attempt to estimate the volumes, which was by Bowen (1967, p. 226). He suggested that:

> Volume calculations based on downwind elliptical distributions . . . indicate that the T-3 event (best known from the Tennessee Carters Limestone) records a minimum ejection of 10 mi^3 [43 km^3] of tephra, whereas the Old Rosedale event (best known from the lower Martinsburg shale of Virginia) records minimum ejection of 24 mi^3 [88km^3].

(I added the metric volumes). The T-3 event refers to the Deicke and the Old Rosedale event to the Millbrig. T-3 is a name used in Kentucky, Tennessee, Georgia, and Alabama for the third K-bentonite from the bottom in the standard sequence described by Wilson (1949). That bed is now identified as the Deicke (Huff and Kolata, 1990). Old Rosedale is a reference to the exposure near Rosedale, Russell County, Virginia, where both the Deicke and Millbrig are completely exposed. Bowen was referring to the K-bentonite at Rosedale that I have identified as the Millbrig (Haynes, 1992) based on its mineralogy and stratigraphic position. At this exposure, the Millbrig, which is actually in the Eggleston Formation and not the lower Martinsburg (Trenton Formation of current usage), is abnormally thick because of shearing and folding, as first noted by Butts (1940).

Obviously there was even more pyroclastic material deposited between the present easternmost exposures of the Deicke and Millbrig in the Blount Group clastic sequence and the actual volcanic vents, but this pyroclastic material and the volcanoes themselves may now be part of an obducted and now deeply eroded terrane of metavolcanic rock in the Piedmont of the eastern U.S., or part of the now deeply buried footwall beneath the Blue Ridge thrust sheet of the southern Appalachians. Regardless of how much additional volume was present initially, it is apparent that both the Deicke and Millbrig ashes were the product of tremendously large volcanic eruptions, each of which was far greater than either the Tambora eruption of 1815 or the Toba eruption of 75 Ka.

Setting of the source volcanoes

Determining the paleogeographic and tectonic setting of the volcanoes from which great volumes of ash were erupted during the Ordovician has been one of the long-standing problems in Appalachian geology since these beds were first described by Nelson (1921, 1922, 1926), and although the problem still exists, significant strides toward an answer have been made in recent years. Samson et al. (1989) suggested that the volcanoes from which these ashes were erupted may have been part of a magmatic arc that developed on continental crust in a convergent tectonic setting. With this scenario it is hypothesized that the Deicke and Millbrig may be genetically related to thick sequences of Ordovician extrusive volcanic rocks in either New Brunswick or the British Isles, rocks whose isotopic ages match those of the Deicke and Millbrig. Rhyodacites and dacites are an important component of those volcanic rocks, so the findings presented herein are compatible with those of Samson et al. (1989) with regard to the petrology of the original Deicke and Millbrig ashes.

Based on the mineralogy, spatial distribution, and internal stratigraphy of the Deicke and Millbrig, a few additional statements can be made about the nature of the eruptions and about the tectonic setting of the vents from which the ash was erupted. Both beds are relatively thin but very widespread, with one or more fining upward sequence, suggesting that the original ashes consisted of great volumes of pyroclastic airfall material rather than pyroclastic flow (ash-flow) deposits. Their volume and lateral extent imply that the ashes were derived from tremendously large eruptions of volcanoes in a convergent setting. These eruptions probably began with a Plinian or ultra-Plinian phase that produced a shard-rich ash, followed by a Peléan phase that produced a crystal-rich ash as a coignimbrite, airfall deposit (W. D. Huff, personal communication). This sequence of events is part of the normal one for volcanic eruptions in which ignimbrite is generated (Sparks et al., 1973). In the various exposures in the eastern Valley and Ridge, the area where the Deicke and Millbrig attain their maximum known thickness and grain size, there is no evidence of near vent deposits in either bed, indicating that even these proximal settings were at a minimum several kilometers from the volcanic vents.

The presence of K-bentonites in much of the Middle and Upper Ordovician of present-day southeastern North America is clear evidence that a volcanic arc was present somewhere along the continental margin. Provenance studies of associated Ordovician sandstones, however, suggest that this arc was not providing sand-sized sediment directly to the associated foreland basin for much of the Ordovician. Mack (1985) found no volcanic rock fragments (VRFs) or other evidence to suggest the presence of a volcanic arc in the source terrane(s) from which the sands of the Bays Formation and other units of the Blount Group in the southern Appalachians were derived. In the central Appalachians farther north, however, Rader and Perry (1976) and Diecchio (1985) identified VRFs in a younger unit, the Upper Ordovician Oswego Formation.

In modern convergent settings the volcanic arc is commonly separated from a foreland basin by a fore-arc basin and associated fore-arc ridge that may consist of an accretionary wedge (Hamilton, 1979). Until the fore-arc ridge is breached, arc-derived sediments will be trapped in the fore-arc basin behind the ridge, and the sand-sized sediments accumulating in the foreland basin will be derived only from erosion of the older passive margin sediments and associated slices of basement rock that make up the accretionary wedge. Mack (1985) indicated that the Blount sediments in the southern Appalachians were derived from just such an assemblage (older sediments and low-grade metasediments with some crystalline input), so accumulation of the Blount Group in a foreland basin shielded from the arc by a fore-arc ridge seems to provide a plausible explanation for Mack's findings. Prior to or during Oswego deposition, the fore-arc ridge must have been breached, and VRFs began entering the foreland basin for the first time.

The provenance of the green facies of the Bays Formation and of the Knobs Formation, key units that are intermediate in age and geographic position to the units in Georgia and Tennessee studied by Mack (1985) and the Oswego of northern Virginia, has not yet been studied in the detail necessary to determine if arc-derived sediments were entering the basin during the time that those units were deposited. Also, no studies on the provenance of mudrocks in the Blount Group have been published, so it is not known whether the muds derived from weathering of the arc were carried by currents over the fore-arc ridge and into the foreland basin. Now underway is a bed-by-bed petrographic and geochemical investigation of the provenance of the sandstones and mudrocks of the green Bays and the Knobs to determine when arc-derived sediments began entering the foreland basin, a finding that would help delimit the overall timing of these tectonically related changes in sediment dispersal patterns. This in turn could help define and limit the paleogeographic and tectonic setting of the volcanoes from which the Deicke and Millbrig were derived. The initial findings of this study indicate that the sands of the Walker Mountain Sandstone Member, the basal unit of the green facies of the Bays, were derived from erosion only of plutonic igneous rocks, low-rank metamorphic rocks, and older sedimentary rocks, with no input from eroding volcanic rocks (Haynes and Goggin, 1993).

Is the Millbrig the "Big Bentonite" of Kinnekulle?

What is probably the best known Ordovician K-bentonite in Scandinavia, the thick "B Bed" or "Big Bentonite" at Kinnekulle, Sweden (Bystrom, 1956; Brusewitz, 1986) has recently been correlated with the Millbrig on the basis of trace element geochemistry, tectonomagmatic setting, and biostratigraphic position (Huff et al., 1992). The Big Bentonite is a biotite- and quartz-rich K-bentonite thicker than 1 m and consisting of three separate graded units (Kunk et al., 1984). These characteristics are strikingly similar to those of the Millbrig at Shelbyville, Tennessee, where three separate graded units are present (Fig. 19). Also, there is very close agreement in the radiometric ages of the two beds, 454 ± 2.1 Ma for the Millbrig and 455 ± 2.0 Ma for the Big Bentonite, obtained from $^{40}Ar/^{39}Ar$ dating of biotites in the Millbrig and biotites and sanidines in the Big Bentonite (Kunk and Sutter, 1984; Kunk et al., 1984). A detailed investigation of the primary plagioclases in the Big Bentonite to determine their compositions would be one step that could help to resolve this question. If they are andesines, like the primary plagioclases in the Millbrig, then another important piece of evidence would favor the trans-Atlantic—and in fact trans-Iapetus—correlation of the Millbrig with the Big Bentonite.

DIAGENESIS

Introduction

The Deicke and Millbrig K-bentonites were originally crystal-rich volcanic ashes that have become crystal-rich potassium bentonites through a sequence of diagenetic changes. Of particular importance in understanding their diagenetic history are the regional variations in the composition of the authigenic feldspars, the mixed-layer I/S clay, and the nonsilicate ferroan and ferrotitanian minerals in the beds. Comparison of compositional trends observed during petrographic study of the Deicke and Millbrig with trends reported in studies of diagenesis in other sedimentary basins provides important clues in understanding the burial history of the K-bentonites and surrounding strata.

Feldspars

In both the Deicke and Millbrig, authigenic albite is present in samples from the southern Valley and Ridge province, where the entire Paleozoic sequence is thousands of meters thick, but authigenic albite is absent in Cincinnati Arch samples, where the Paleozoic sequence is much thinner. Huff and Türkmenoğlu (1981) noted abundant K-feldspar in some K-bentonite samples from along the Cincinnati Arch, but they, too, did not observe any albite. Authigenic K-feldspar occurs in both beds throughout most of the study area.

Table 5 compares the feldspar mineralogy of selected Deicke and Millbrig samples with the conodont color alteration index (CAI) values for Ordovician strata from the same localities (Harris et al., 1978). Because CAI values are directly related to temperature (Epstein et al., 1977), this table shows that the occurrence of authigenic albite is restricted to those areas exposed to higher burial temperatures. Harris et al. (1978) suggested that the thermal gradients delimited by the CAI values were the result of regional variations in overburden thickness.

The authigenic albite and authigenic K-feldspar replace primary plagioclase grains (Figs. 6D and 7E), and albite replaces authigenic K-feldspar as well (Figs. 6A and 7D). The geochemistry of these and related reactions has been discussed by Kastner and Siever (1979), Fisher and Land (1986), Haynes (1989), and Morad et al. (1990), and the temperature dependence of feldspar authigenesis has been demonstrated by several investigations of feldspathization in sedimentary basins. Albitization begins to occur between 75° and 120°C, with the most common range being from 100° to 120°C, and at lower temperatures albite authigenesis does not occur (Land and Milliken, 1981; Boles, 1982; Walker, 1984; Fisher and Land, 1986; Gold, 1987; Morad et al., 1990). These studies also show

TABLE 5. COMPARISON OF CONODONT CAI VALUES* WITH THE FELDSPAR MINERALOGY OF SELECTED DEICKE AND MILLBRIG SAMPLES

Sample†	Formation	Conodont CAI Value§	Authigenic Feldspars	Intermediate Plagioclases
Deicke: Cincinnati Arch sections				
BB 5-5-1	Tyrone	≤1.5	K-Feldspar	Labradorite
BRM T-3	Tyrone	≤1.5	K-Feldspar	Labradorite
TN 1-2	Tyrone	≤1.5	K-Feldspar	Labradorite
Deicke: Valley and Ridge carbonate sections				
AL 1	Chickamauga (Stones River)	2–2.5	K-Feldspar and albite	Labradorite
AL 7	Stones River	3	K-Feldspar and albite	Labradorite
GA:DD 2-1	Carters	2.5–3	K-Feldspar	None present
GA:WL 3-1	Carters	2.5–3	K-Feldspar and albite	Labradorite
TN 12-1	Carters	1.5–2	K-Feldspar	Labradorite
TN 10	Chickamauga (Eggleston)	1.5–2	K-Feldspar and albite	Labradorite
TN:CL 1-1	Eggleston	1.5–2	K-Feldspar	Indeterminable
VA 2-1	Eggleston	2	K-Feldspar and albite	Labradorite
VA:SC 1-1	Moccasin	2.5	K-Feldspar and albite	Labradorite
VA:TZ 1-7	Moccasin	2.5–3	K-Feldspar	Indeterminable
Deicke: Valley and Ridge sandstone sections				
AL:CH 1-2	Colvin Mountain	3–3.5	None present	None present
AL:ET 1-2	Colvin Mountain	3–3.5	None present	None present
GA:FL 1-1	Colvin Mountain	3–3.5	K-Feldspar	None present
Millbrig: Cincinnati Arch sections				
BRM T-4	Tyrone	≤1.5	K-Feldspar	Andesine
KRI 5-0	Tyrone/Lexington contact	≤1.5	K-Feldspar	Andesine
TN 7-4	Carters/Hermitage contact	≤1.5	K-Feldspar	Andesine
Millbrig: Valley and Ridge carbonate sections				
AL 3-4	Chickamauga (Stones River)	2–2.5	K-Feldspar and albite	Andesine
AL 9-1	Stones River	3	K-Feldspar and albite	Andesine
AL 17	Stones River	2.5–3	K-Feldspar and albite	Andesine
TN 11-1	Chickamauga (Eggleston)	1.5–2	K-Feldspar and albite	Andesine
VA 4-1	Eggleston	2	K-Feldspar and albite	Andesine
VA:SC 1-2	Eggleston	2.5	K-Feldspar and albite	Andesine
VA:GI 1-1	Eggleston	3.5–4	Indeterminable	Indeterminable
Millbrig: Valley and Ridge sandstone sections				
AL:CH 1-4	Colvin Mountain	3–3.5	None present	None present
TN:HK 2-1	Bays	3–3.5	Indeterminable	Indeterminable
VA:SM 1-4	Bays	3–3.5	Indeterminable	Indeterminable
VA:WY 2-1	Bays	3.5–4	K-Feldspar	Indeterminable
VA:WY 1-1	Bays	3.5–4	Indeterminable	Indeterminable
VA:RO 1-3	Bays	4–4.5	Indeterminable	Indeterminable
VA:BT 2-1	Bays	4–4.5	K-Feldspar	Indeterminable

*Harris et al., 1978.
†Sample localities are given in Appendix 1.
§The location of many of the exposures from which K-bentonites were collected are shown precisely on the map of Harris et al. (1978), and the CAI value was easily obtained. CAI values for exposures not shown were determined as closely as possible using an appropriately scaled overlay.

that authigenic albite remains stable at and above the range of temperatures in which dissolution of K-feldspar and primary intermediate plagioclase tends to occur. Authigenesis of K-feldspar, however, readily occurs in sedimentary and volcanic/volcaniclastic rocks under near surface conditions (Hay, 1962; Walker et al., 1978; Waugh, 1978; Ali and Turner, 1982; Marshall et al., 1986). Significantly, Kastner and Siever (1979) found that the geochemical environment favoring albitization does not need to differ appreciably from the geochemical environment favoring authigenesis of K-feldspar; however, albite authigenesis is favored thermodynamically at higher temperatures, given a suitable pore-water geochemistry.

The lack of authigenic albite in samples from the Cincinnati Arch probably indicates that the maximum temperature experienced by the K-bentonites in that region was less than 75°C; however, the maximum temperature experienced by the beds in the southern Valley and Ridge was, at a *minimum,* probably around 110°C. As a result, authigenic albite formed throughout that region and is present today in samples of the K-bentonites from the southern Valley and Ridge province, in striking contrast to its complete absence from Cincinnati Arch samples.

Clay minerals

It is widely accepted that I/S in altered volcanic ashes forms as the end product of a process that starts when felsic volcanic glass devitrifies and smectite forms. Over time this smectite is transformed to I/S as aluminum substitutes for silicon in tetrahedral coordination, and potassium is taken up at selected locations in the interlayer position (Środoń and Eberl, 1984). In sedimentary basins, both the illite:smectite ratio and the degree of ordering of the I/S (a function of that ratio) are dependent on temperature (and thus depth of burial), so that as diagenesis becomes low-grade metamorphism the increase in temperature causes smectite to become I/S, then a well-ordered illite, and eventually muscovite (Maxwell and Hower, 1967; Perry and Hower, 1970; Weaver and Beck, 1971; Boles and Franks, 1979; Hoffman and Hower, 1979; Dypvik, 1983). This process is kinetically as well as thermodynamically dependent (Elliott et al., 1991).

The ordering of the I/S, and thus the ratio of illite to smectite, tends to increase in Deicke and Millbrig samples from the Cincinnati Arch toward the south and southeast into the southern Valley and Ridge province, and along strike in the Valley and Ridge from Alabama to Virginia (Figs. 12 and 13). Changes in CAI values show that the trend of increasing temperatures in Ordovician strata was from northwest to southeast across the entire Appalachian basin during the Middle and Late Paleozoic, and also from southwest to northeast within the Valley and Ridge portion of the basin (Harris et al., 1978). Table 6 compares the ordering and the percentage of illite in the I/S of selected Deicke and Millbrig samples with the conodont CAI values of Ordovician carbonate samples collected from the same localities by Harris et al. (1978). The variations in ordering and percentage of illite suggest increasing temperatures from northwest to southeast, and from Alabama to Virginia, a trend that is in agreement with that indicated by CAI values in the region.

Fe and Fe-Ti minerals

The feldspars and clay minerals clearly are the most useful indicators of the submetamorphic diagenetic changes that have taken place in the Deicke and Millbrig with regard to variations in temperature. The regional distribution of the nonsilicate ferroan and ferrotitanian minerals is also related to the diagenetic history of the beds, specifically the redox potentials of the diagenetic environment. Those minerals are most common in the Deicke, and the regional variations in their composition are given in Table 7.

Diagenesis of nonsilicate ferrotitanian minerals in sedimentary sequences is discussed by Morad and AlDahan (1986) and Morad (1988b), and many of the inferences concerning diagenesis of the ferroan and ferrotitanian minerals in the Deicke and Millbrig K-bentonites and adjacent strata are based on those publications. The distributions of these minerals in the Deicke (Table 7) are not random. Figure 20 shows how these regional variations can be readily grouped by the classification chart of Maynard (1982). This chart is used to determine the particular diagenetic environment of Berner (1981) that had the most lasting effect on a given sedimentary sequence. Figure 21 shows the areal distribution of these regional variations.

The major ferroan mineral in the primary phenocryst assemblage of the Deicke was ilmenite/titanomagnetite, and it has altered to pyrite and TiO_2 minerals throughout the Cincinnati Arch and much of the western Valley and Ridge (Figs. 6C, E, and 9A, B), areas where the K-bentonites occur in gray limestones that are commonly pyritic (Figs. 20 and 21, Area A). Alteration of ilmenite to pyrite and TiO_2 minerals occurs under reducing conditions where sulfur is present in pore waters as sulfide species (Morad and AlDahan, 1986). The pyrite in the Deicke and the surrounding limestones of Area A is an indication that sulfides were available to form pyrite by combining with the Fe^{2+} ions freed from the ilmenite. This process occurred in the anoxic sulfidic environment of Berner (1981), the diagenetic environment that had the most effect on Area A.

Samples of the Deicke from exposures of the Colvin Mountain Sandstone (Figs. 20 and 21, Area D) contain ilmenite in moderate abundance, but little else except the clay minerals I/S and kaolinite (Fig. 12). This ilmenite is the only ferroan mineral in the Deicke; in fact, the only primary phenocrysts of any kind in either the Deicke or the Millbrig from Colvin Mountain Sandstone exposures are these grains of unaltered to slightly altered ilmenite (Fig. 9D). The Colvin Mountain Sandstone consists mainly of compositionally and texturally mature quartz sand that was reworked and eventually deposited in a well-oxygenated nearshore setting. These sands are bounded below and above by hematitic redbed sequences, the Greensport and Sequatchie Formations, respectively (Jenkins, 1984). There is no pyrite in any of these units.

Ilmenite is very stable under oxidizing conditions, and what alteration does occur produces the various ferrotitanian phases that are intermediates between ilmenite and leucoxene (Morad and AlDahan, 1986). Hematite is also stable under oxidizing conditions; pyrite, of course, is not. Thus it is inferred that during diagenesis the pore waters of the Colvin Mountain sands and adjacent sediments were never reducing, given the

TABLE 6. COMPARISON OF CONODONT CAI VALUES* WITH THE CLAY MINERALOGY OF SELECTED DEICKE AND MILLBRIG SAMPLES

Sample†	Formation	Conodont CAI Value§	Average Illite** (%)	Ordering**
Deicke: Cincinnati Arch sections				
BB 5-5-1	Tyrone	≤1.5	77	R1
Deicke: Valley and Ridge carbonate sections				
AL 7	Stones River	3	87	R1/R3
GA:DD 2-1	Carters	2.5–3	87	R1/R3
TN:CL 1-1	Eggleston	1.5–2	92	R3
VA:SC 1-1	Moccasin	2.5	92	R3
VA:RL 2-1	Eggleston	3.5	83	R1/R3
VA:TZ 1-7	Moccasin	2.5–3	90	R3
VA:BL 1-1	Eggleston	2.5–3	92	R3
VA:GI 4-0	Moccasin	3.5–4	93	R3
Deicke: Valley and Ridge sandstone sections				
AL:CH 1-2	Colvin Mountain	3–3.5	85	R1/R3
AL:ET 1-2	Colvin Mountain	3–3.5	98	R3
GA:FL 1-1	Colvin Mountain	3–3.5	95	R3
Millbrig: Cincinnati Arch sections				
BB 5-2-3	Tyrone	≤1.5	80	R1
KRI 5-0	Tyrone/Lexington contact	≤1.5	76	R1
TN 7-3	Carters/Hermitage contact	≤1.5	75	R1
TN 7-4	Carters/Hermitage contact	≤1.5	79	R1
Millbrig: Valley and Ridge carbonate sections				
GA:DD 2-2	Carters	2.5–3	76	R1
GA:WL 3-2	Carters	2.5–3	87	R3
TN:CL 1-2	Eggleston	1.5–2	88	R1/R3
VA 4-1	Eggleston	2	86	R1/R3
VA:LE 1-2	Eggleston	2–2.5	87	R3
VA:SC 1-2	Eggleston	2.5	90	R3
VA:GI 1-1	Eggleston	3.5–4	94	R3
Millbrig: Valley and Ridge sandstone sections				
AL:CH 1-4	Colvin Mountain	3–3.5	80	R1/R3
VA:SM 1-1	Bays	3–3.5	88	R3
VA:WY 2-1	Bays	3.5–4	88	R3
VA:WY 1-1	Bays	3.5–4	97	R3
VA:BT 2-1	Bays	4–4.5	89	R3

*Harris et al., 1978.
†Sample localities are given in Appendix 1.
§The location of many of the exposures from which K-bentonites were collected are shown precisely on the map of Harris et al. (1978), and the CAI value was easily obtained. CAI values for exposures not shown were determined as closely as possible using an appropriately scaled overlay.
**The average illite percentage and the ordering were calculated from x-ray diffraction tracings using the technique of Watanabe (1981), summarized by Srodon and Eberl (1984).

lack of pyrite and presence of hematite in the sediments and the presence of unaltered to only slightly altered ilmenite phenocrysts in the Deicke. This mineralogy indicates that diagenesis of the Colvin Mountain Sandstone occurred in the oxic environment of Berner (1981). Evidently this oxic environment produced pore waters that were chemically very aggressive, as indicated by (1) the total absence of any phenocrysts (including quartz) other than ilmenite in either the Deicke or Millbrig at the Greensport Gap and Alexander Gap sections, (2) the significant kaolinization of the K-feldspars that are present in the Deicke from the Horseleg Mountain section, and (3) the presence of appreciable kaolinite in the <2-μm size fraction of all Deicke and Millbrig samples from exposures of the Colvin Mountain Sandstone. Together, these findings indicate that the sediments have been very strongly leached.

Where the Deicke occurs in the Moccasin Formation in

TABLE 7. DISTRIBUTION OF NONSILICATE Fe AND Fe-Ti MINERALS IN THE DEICKE BASED ON THIN SECTION ANALYSES OF SELECTED SAMPLES

Sample*	Formation	Fe-Ti Minerals Present†
Cincinnati Arch sections		
BB 5-5-1	Tyrone	Pyrite, TiO_2 polymorphs
N-5	Tyrone	Pyrite, TiO_2 polymorphs
BRM T-3	Tyrone	Pyrite, TiO_2 polymorphs
KY 6-1	Tyrone	Pyrite, magnetite, TiO_2 polymorphs
C-114 T-3	Tyrone	Pyrite, TiO_2 polymorphs
C-113 T-3	Tyrone	Pyrite, TiO_2 polymorphs
KY 1-3	Tyrone	Pyrite, TiO_2 polymorphs
TN 3-3	Carters	Pyrite, magnetite, TiO_2 polymorphs
CHT-3	Carters	Pyrite, TiO_2 polymorphs
TN 1-2	Carters	Pyrite, TiO_2 polymorphs
Valley and Ridge carbonate sections		
AL 1	Chickamauga (Stones River)	Pyrite
AL 7	Stones River	Pyrite, TiO_2 polymorphs
AL:ET 2-1	Stones River	Pyrite
AL 12-1	Stones River	Pyrite
GA:DD 1-1	Carters	Pyrite, TiO_2 polymorphs
GA:DD 2-1	Carters	Pyrite
GA:WL 3-1	Carters	Pyrite, TiO_2 polymorphs
TN:HM 2-1	Carters	Pyrite, TiO_2 polymorphs
TN 10	Chickamauga (Eggleston)	Pyrite, TiO_2 polymorphs
TN 12-1	Carters	Pyrite, TiO_2 polymorphs
TN:CL 1-1	Eggleston	Pyrite
TN 14-2	Moccasin	Hematite
TN:HK 3-1	Moccasin	Hematite
VA 2-1	Eggleston	Pyrite, TiO_2 polymorphs
VA:LE 1-1	Eggleston	Pyrite
VA:SC 1-1	Moccasin	Hematite
VA:RL 2-1	Eggleston	Hematite
VA:TZ 1-7	Moccasin	Hematite
VA:TZ 2-1	Moccasin	Hematite
VA:BL 1-1	Eggleston	Hematite
WV:ME 1-1	Moccasin	Hematite
VA:GI 4-0	Moccasin	Hematite
Valley and Ridge sandstone sections		
AL:CH 1-2	Colvin Mountain	Ilmenite, TiO_2 polymorphs
AL:ET 1-2	Colvin Mountain	Ilmenite, TiO_2 polymorphs
GA:FL 1-1	Colvin Mountain	Ilmenite

*Sample localities are given in Appendix 1.
†TiO_2 polymorphs are anatase or rutile, and possibly brookite, and in the Colvin Mountain Sandstone, leucoxene.

Virginia, West Virginia, and Tennessee, the only nonsilicate ferroan mineral present is hematite. The presence of hematite and the lack of ilmenite is a major difference between Deicke samples from the oxidized sediments of the Moccasin and Deicke samples from the apparently equally oxidized sediments of the Colvin Mountain (many geologists have in fact noted the striking textural and compositional similarities between the Moccasin Formation and the Greensport Formation, which immediately underlies the Colvin Mountain).

Because ilmenite is readily altered to pyrite and TiO_2 minerals under reducing conditions (Morad and AlDahan, 1986), the lack of ilmenite in samples of the Deicke from the Moccasin suggests that early diagenesis occurred in the anoxic sulfidic environment (Figs. 20 and 21, Area C). This diagenesis would have to have been early rather than late because the abundant hematite now disseminated in the Moccasin is evidence that later diagenesis occurred in the oxic environment.

Such a period of early diagenesis most likely occurred subsequent to deposition and prior to deep burial. The Moccasin is interpreted as a sequence of peritidal mudflats (Hergenroder, 1966; Kreisa, 1980; Simonson, 1985), and in peritidal muddy sediments on modern mudflats all but the upper several millimeters are commonly anoxic, even when the overlying water column is well oxygenated. Development of anoxic conditions in modern mudflat sediments can occur even when they are subaerially exposed for significant periods of time (Maynard, 1982), as parts of the Moccasin appear to have been (Kreisa, 1980; Simonson, 1985), given the presence of desiccation cracks and other exposure indicators. The rapid depletion of oxygen just beneath the sediment/air or sediment/water interface is a process that can be especially noticeable in muddy, fine-grained sediments (Maynard, 1982), like those of the Moccasin.

Later diagenesis of the Moccasin Formation occurred in an oxic environment, and at this time the pyrite and other minerals containing ferrous iron were oxidized. It was this oxic environment that clearly had the more lasting effect on the Moccasin sediments and on the Deicke itself. Unlike the gray limestones farther west, which contain pyrite and organic matter and are clearly not oxidized except from surface weathering, the sediments in the Moccasin have been thoroughly and pervasively oxidized, and all iron is present as disseminated hematite, which gives the Moccasin its distinctive red color. Hematite and gypsum, which occur in the Moccasin, are the alteration products of pyrite in an oxic environment, although oxidation of the pyrite in the Deicke could only have been a very minor contributor overall of iron and sulfur to the sediments.

Although the timing of the oxidation of the Moccasin sediments is not well constrained, it is likely that appreciable oxidation of the pyrite in the Deicke and the surrounding sediments of the Moccasin occurred when the significant volumes of upland-sourced meteoric waters moved into and through the sedimentary sequence along the southeastern Valley and Ridge and flushed out the connate, reducing pore waters that had remained in the sediments throughout much of the region at that time (Grover and Read, 1983). This event occurred during the Late Ordovician or Silurian, probably continuing until the time that the widespread pre-Devonian unconformity developed. Those oxidizing meteoric waters, the surface drainage from the tectonic highlands then in existence to the southeast (present-day), were responsible for the distinctive zonation in the carbonate cements of some of the limestones that underlie the Moccasin and the Bays Formations (Grover and Read, 1983). Presumably, those waters would have moved

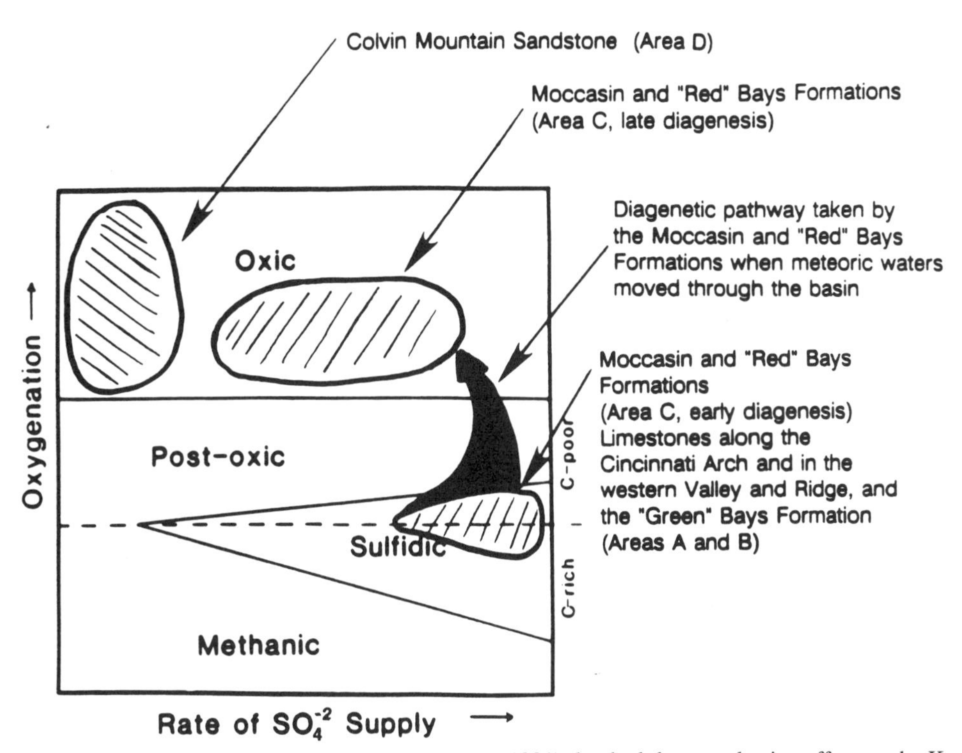

Figure 20. The diagenetic environments of Berner (1981) that had the most lasting effect on the K-bentonites and the surrounding sediments in the four areas shown in Figure 21. The fields for Areas A and B and for early diagenesis of Area C straddle the C-rich and C-poor line in the anoxic sulfidic environment because field evidence suggests that in some areas anoxic conditions existed at the base of the water column as well as in the sediment, (no bioturbation), whereas in other areas anoxic conditions were present only beneath the sediment-water interface (limited bioturbation present). Diagenesis of Area D and the late diagenesis of Area C occurred in the oxic environment. Base diagram from Maynard (1982).

through what are now the redbeds of the Moccasin and Bays Formations as well. Eventually, these oxidizing waters stagnated, probably in the later Silurian, and redox conditions in the pore waters gradually became reducing again basinwide. This change, the existence of which is shown by the youngest zones of carbonate cements in the underlying units (Grover and Read, 1983), appears to have had little discernable effect on the diagenesis of the Moccasin and the Deicke. Study of the later generation of carbonate cements in fenestrae of the Moccasin limestones would be helpful in determining whether the late stage cementation history of the underlying limestones described by Grover and Read (1983) can be detected in the Moccasin as well.

The Deicke does not occur in the Bays Formation of Virginia and northeast Tennessee, so inferences about diagenesis must be based on a consideration of the ferroan minerals in the Millbrig and in the sediments of the Bays itself. In Tennessee and most of southwestern Virginia, the Bays is a sequence of red, fine-grained calcareous mudrocks of tidal flat origin. Its texture and composition are similar to the mudrocks of the Moccasin Formation, but the Bays contains coarser sandstones not present in the Moccasin. This is the red facies of the Bays Formation (Haynes, 1992). That formation, like the Moccasin, is colored by disseminated hematite. In the Salem synclinorium and adjacent areas near Roanoke, it is interbedded with, and largely replaced by, a sequence of coarser grained greenish gray noncalcareous lithic sands deposited in a delta front setting. This is the green facies of the Bays Formation (Haynes, 1992).

Where the Millbrig occurs in the red Bays, the biotites have been extensively chloritized and kaolinized (Fig. 10D), and the clay matrix is red, just like the sediments of the Bays itself. In the green Bays, however, the Millbrig is much less altered, and samples of it are very similar texturally to samples from the gray limestones farther west in the Valley and Ridge and the Cincinnati Arch; the biotites are dark bronze or brown to black, and the clay matrix is greenish gray. In addition, pyrite or its oxidized successor limonite, occurs in the siltstones and sublithic arenites of the green Bays (Hergenroder, 1966).

Like the Moccasin the red Bays was most affected by diagenesis in an oxic environment. By contrast, the pyritic green Bays was not oxidized during the mid-Paleozoic when the meteoric waters draining the tectonic highlands to the southeast (present-day) came into the basin (Grover and Read, 1983) and

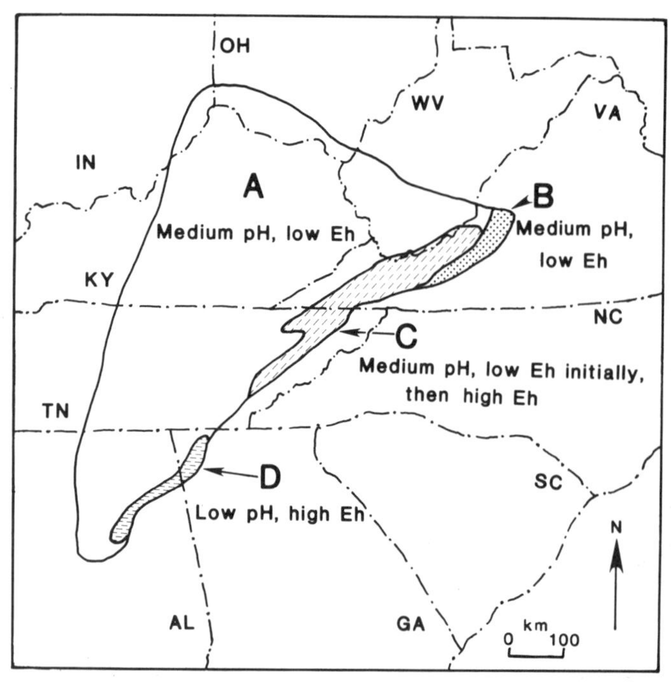

Figure 21. Regional variations in Eh and pH of the diagenetic environments that had the most significant effect on the mineralogy of the Deicke and Millbrig K-bentonites and the adjacent sediments. Region A includes exposures of the Deicke and Millbrig in the Tyrone and Carters Limestones along the Cincinnati Arch, and the Eggleston Formation, Carters Limestone, Stones River Group, and Chickamauga Limestone in the western Valley and Ridge. Region B includes exposures of the Millbrig in the green Bays Formation and equivalent beds in the Salem and Fincastle synclinoria and at the Connor Valley and Crockett Cove sections in the eastern Valley and Ridge of Virginia. Region C includes exposures of the Deicke and Millbrig in the red Bays, Moccasin, and Eggleston Formations in the eastern and central belts of the Virginia and Tennessee Valley and Ridge. Region D includes exposures of the Deicke and Millbrig in the Colvin Mountain Sandstone of the Alabama and Georgia Valley and Ridge.

are thought to have oxidized the iron in the Moccasin and the red Bays farther south. Instead, the green Bays and the adjacent gray limestones of the Salem synclinorium and areas to the north and northwest were most affected by diagenesis in the anoxic sulfidic environment. Thus the diagenetic history of the red Bays (Area C) parallels that of the Moccasin Formation (also Area C), whereas the diagenetic history of the green Bays (Area B) more closely parallels that of the pyritic limestones of the western Valley and Ridge and the Cincinnati Arch (Area A), as shown in Figure 20.

Unlike the pyrite in the Deicke, the pyrite in the green Bays is presumably not derived from alteration of ilmenite, nor is it the result of very slow sedimentation rates like the pyrite that is commonly present in condensed sequences deposited in anoxic environments. Instead, it probably formed from the reaction of labile ferroan minerals with organic matter during the rapid burial of these delta front green Bays sediments. As noted by Maynard (1982), the rate of sulfate reduction, and therefore the likelihood that pyrite will be present in a sediment, tends to increase with increasing sedimentation rates. This phenomenon adequately explains the presence of pyrite in the green Bays Formation of the Salem synclinorium and nearby areas. The few red beds in the upper green Bays in some sections (Hergenroder, 1966; Kreisa, 1980) may have been oxidized during localized shallowing and slowing of sedimentation rates, or, given the proximity at some sections of fault zones that could have acted as pathways for oxidizing meteoric waters during the tectonic activity that occurred in this region during the later Paleozoic, the red coloration could be a later oxidation event unrelated to depositional and diagenetic environments. The interbedding of pyritic (usually green) and hematitic (usually red) sediments is a common but geochemically enigmatic occurrence in many sedimentary sequences, including several in this region such as the Lower and Middle Cambrian Waynesboro Formation (Haynes, 1991), the Upper Ordovician Oswego and Juniata Formations (Diecchio, 1985), and the Upper Devonian Foreknobs and Hampshire Formations (Woodrow et al., 1988) The differences in color are typically attributed to variations in the redox conditions of the depositional environment, but the complex diagenetic history of the Moccasin Formation and the Deicke K-bentonite suggests that the explanation is not always that simple.

Samson and others (1988) analyzed Deicke samples from the Upper Mississippi Valley and T-3 (Deicke) samples from the Cincinnati Arch, and they concluded that the Deicke once contained biotite, which has since altered to TiO_2 and ilmenite. Based on the findings of the present study this conclusion seems unlikely because: (1) both TiO_2 and biotite are present in the Deicke (Table 1); (2) the Deicke TiO_2 minerals are pseudomorphic after ilmenite, not biotite; and (3) ilmenite in the Deicke is a primary phenocryst, not a secondary mineral. There is no indication that the many ferrotitanian minerals in the Deicke are diagenetically related in any way to the few biotite grains usually present. In the least weathered samples studied (from cores on the Cumberland Saddle of Kentucky and Tennessee; Fig. 1), the few biotites in the Deicke are little altered but all the ilmenite grains are partly to completely altered to TiO_2 minerals (Figs. 6C and 9A, B). Furthermore, there is no evidence that the diagenetic environment differed appreciably in the few meters that separate the Deicke and Millbrig. Throughout the region where the T-3 (Deicke) samples of Samson et al. (1988) were collected, the Millbrig is only 3 to 8 m upsection from the Deicke. TiO_2 minerals are not found in the Millbrig, and even more significantly, the iron and titanium in the Millbrig are still part of the biotite crystal structure in most samples. Because there is no ilmenite in the Millbrig, a significant source of titanium other than the biotite is unlikely.

This evidence seems to rule out the possibility that biotites in the Deicke (of which there are few), but not those in the Millbrig (of which there are many), could selectively have been altered to ilmenite or TiO_2 minerals (of which there are many in the Deicke), and at the same time the numerous biotites in the Millbrig remained essentially unaltered.

It is not surprising that Samson et al. (1988) found no biotite in Deicke samples from the Upper Mississippi Valley. Even Millbrig samples from that area contain vanishingly small amounts of biotite, and both the Deicke and Millbrig contain few phenocrysts, being composed instead primarily of mixed-layer I/S (Kolata et al., 1986), the product of altered volcanic glass. In exposures along the Cincinnati Arch and farther east in the southern Valley and Ridge, both the Deicke and Millbrig contain abundant nonclay minerals, including the abundant biotite in the Millbrig. Although present, biotite is never abundant in the Deicke. Therefore, the presence of more than a few grains of biotite in several kilograms of sample would be unexpected from samples of the Deicke in the Upper Mississippi Valley sections.

Discussion

The observed regional differences in authigenic feldspar compositions and in the illite:smectite ratio and the ordering of the I/S in the Deicke and Millbrig are the result of variations in thermal gradients, presumably the same that produced the regional differences in conodont CAI values (Harris et al., 1978), as there is reasonably good agreement between the regional variations observed for these parameters (Tables 5, 6). These gradients are likely related to variations in the thickness of post-Ordovician overburden throughout the southern Appalachians. As the depth of burial increased throughout the mid-Paleozoic, the temperature of the pore waters in the ever more deeply buried sediments rose as well. Once these pore waters were mobilized following dewatering of the Sevier and other shale basins during the Late Paleozoic (Oliver, 1986; Kesler et al., 1988), the temperature of the pore waters in a given region would be a function of the existing thermal gradients caused by differences in overburden thicknesses.

Migration in various directions of these saline pore waters through the Appalachian basin sedimentary fill is thought to have caused the formation of authigenic feldspar in Cambrian limestones (Hearn and Sutter, 1985), the illitization of the Ordovician Deicke and Millbrig K-bentonites and certain Devonian K-bentonites (Elliott and Aronson, 1987, 1993), and the formation of Mississippi Valley–type lead-zinc and fluorspar deposits in various strata (Duane and de Wit, 1988). This brine migration occurred subsequent to the influx of meteoric waters into the southeastern edge of the Appalachian basin, based on the best estimates of the timing of both events (Grover and Read, 1983).

Conodont CAI values (Harris et al., 1978) and the regional variations in feldspar and clay mineralogy in the Deicke and Millbrig K-bentonites indicate that the Valley and Ridge province experienced higher burial temperatures than did the Cincinnati Arch, and that in the southern Valley and Ridge the highest temperatures were in southwestern Virginia. Therefore, the observed variations in the illite:smectite ratio in the I/S of the K-bentonites support the migrating brine theory, as does the distribution of authigenic K-feldspar and albite, because the extent to which the illitization reaction occurs is dependent on the temperature of the brine at the time of the reaction (Morton, 1985), just as the formation of authigenic albite versus authigenic K-feldspar is governed by temperature (Kastner and Siever, 1979). Albite would have formed preferentially instead of K-feldspar in the warmer pore waters that existed in the more deeply buried Valley and Ridge area, and only K-feldspar would have formed in the more shallowly buried Cincinnati Arch area, where the brines would have been cooler. The temperature of the brines would presumably have reflected the prevailing burial temperatures and, by the time they reached the Cincinnati Arch, they would have cooled appreciably by conduction.

Migrating brines probably had relatively little effect on the diagenesis of the nonsilicate ferroan minerals in the Deicke. Those minerals were instead more affected by changes in the redox potentials of the pore waters throughout the basin, and in much of the region they probably began to alter soon after deposition.

The observed regional variations in the feldspar and clay mineralogy of the K-bentonites do not answer all the questions about the burial history of the Deicke and Millbrig; three questions remain unsolved:

1. The two-feldspar grains occur in the Deicke and Millbrig throughout the southern Valley and Ridge, from west-central Virginia to central Alabama, and the distance traveled by the expelled brines from the Sevier basin is unknown. Could the authigenic feldspars observed in Alabama have been formed by the older, cooler brine discussed by Kesler et al. (1988), or are they from the younger, hotter brine? Or are they the result of brines derived from the dewatering of other shales, for example, those in the Rome Formation or the Conasauga Group (Cambrian), or the Chattanooga/Millboro Shale (Devonian), and therefore completely unrelated to the brines described by Kesler et al. (1988) that came from the dewatering of the Sevier basin? (This includes the basin in which the Liberty Hall Formation, Paperville Shale, Rich Valley Formation, and other Sevier equivalents were deposited in Virginia.)

2. Are the authigenic K-feldspars in K-bentonite samples from the Cincinnati Arch significantly younger than those in samples from the western Valley and Ridge, as the migrating brine theory would predict? And what are the ages of the K-feldspars in the Valley and Ridge relative to those in the surrounding sediments such as the feldspars studied by Hearn and Sutter (1985)? Obviously, absolute ages of the authigenic K-feldspars will have to be determined if those and other questions concerning the formation of the authigenic feldspars are to be answered.

3. The data of Elliott and Aronson (1987, 1993) show that illitization of both Ordovician and Devonian K-bentonites occurred in the Late Paleozoic. If authigenic feldspars were forming 100 m.y. prior to the illitization, as seems possible

based on the evidence of Kesler et al. (1988), what is the connection between the two episodes of authigenesis? The ages of illitization reported by Elliott and Aronson (1987, 1993) agree with ages for authigenic K-feldspar in Cambrian carbonates of western Maryland reported by Hearn and Sutter (1985), suggesting a complex history of feldspar authigenesis and illitization throughout the Middle and Late Paleozoic in the Appalachian basin. For example, it is possible that at some localities the authigenic feldspars in the Deicke and Millbrig were formed at some time between 450 and 310 Ma, but elsewhere in the basin the feldspathization may have occurred over a relatively short period of time, about 310 to 295 Ma. The older interval of time is when many of the textures observed by Kesler et al. (1988) were formed, and the younger interval of time is when illitization of the smectite in the Deicke and Millbrig occurred (Elliott and Aronson, 1987) and also when K-feldspar authigenesis occurred in the Cambrian carbonates of Maryland (Hearn and Sutter, 1985). Therefore, it is not possible to determine at this time whether feldspathization occurred during one relatively brief period of time, the favored mechanism for illitization of the beds (Morton, 1985), or whether there were intermittent episodes of feldspathization throughout the Middle and Late Paleozoic. More K/Ar ages of the I/S will need to be obtained from additional Deicke and Millbrig samples to determine if the illitization episode is uniform in age throughout the southern Valley and Ridge, as the precise but geographically limited data of Elliott and Aronson (1987, 1993) suggest. Unfortunately, conodont colors cannot help to resolve the timing of either feldspathization or illitization, because their color is dependent only on temperature, not chemical composition, and the time at which the color change occurred cannot be determined.

Summary of burial history

The regional variations in feldspars and clays in the Deicke and Millbrig suggest that Ordovician strata of the Cincinnati Arch were heated to a maximum temperature of between 75° and 100°C, but in the southern Valley and Ridge province they were heated to a minimum temperature of about 110°C. These temperatures correspond to post-Ordovician burial depths of no more than about 2,100 m along the Arch, and about 2,500 m to perhaps more than 4,000 m in the Valley and Ridge, and they agree very well with the thermal maturation data of Harris et al. (1978) that are based on conodont CAI values. The CAI values, however, are more sensitive to incremental changes in temperature between 50° and 100°C, unlike the reactions governing K-feldspar and albite authigenesis, which are sensitive only to a single temperature change, around 100°C (Kastner and Siever, 1979). Therefore, maximum burial depths along the Cincinnati Arch and in the Valley and Ridge are best defined using conodont CAI values. For example, the conodont CAI values for samples from the Cincinnati Arch indicate that the maximum temperature in that region was about 50°C, a temperature that implies a maximum burial depth of about 1,500 m (Harris et al., 1978) rather than the maximum 2,100 m that is indicated by the feldspar distributions in the K-bentonites.

The suggested burial paths taken by the strata in which the Deicke and Millbrig occur are shown in Figure 22 for different parts of the southern Appalachian basin. The percentage of illite in the I/S of the K-bentonites and the changes in conodont CAI values (Harris et al., 1978) are also shown for comparison. Probably up to the Late Devonian or Early Mississippian, the burial depth of the Deicke and the Millbrig ranged from less than 1,000 m up to 3,000 m basinwide. During that time, the original felsic volcanic glass devitrified, forming smectite, and perhaps authigenic K-feldspar as well. This feldspathization process, which resulted in the partial or complete replacement of primary andesine or labradorite (and perhaps primary sanidine) in both the Cincinnati Arch and Valley and Ridge, may have begun when the overburden was only a few hundred meters thick, as Hay (1962) and Waugh (1978) have shown possible. Depending on the temperature

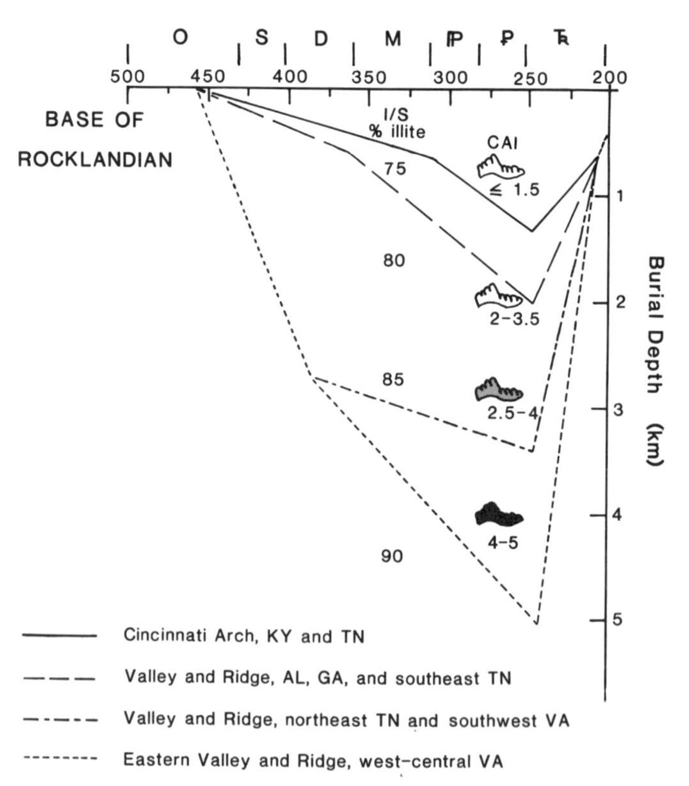

Figure 22. Suggested burial curves from the Ordovician into the Triassic for the Deicke and Millbrig K-bentonites in the southeastern United States. The four separate curves reflect the differences in burial history of Ordovician strata across the Appalachian basin. The changes that occur in I/S mineralogy and conodont CAI values across the basin are shown together for comparison with each other. These curves are based on the data in Tables 4, 5, and 6 of the present study, and the work of Harris et al. (1978).

and composition of brines being expelled from the Sevier basin or elsewhere during that time (Kesler et al., 1988), albite may also have been forming in parts of the southern Valley and Ridge. In the reducing environment that was prevalent the ilmenite in the Deicke began to alter to pyrite except in the Colvin Mountain Sandstone of Alabama and Georgia, where oxidizing conditions persisted following burial.

From the Late Ordovician until perhaps the Late Devonian, temperatures would have been between 20° and 60°C, assuming a geothermal gradient of 2.5°C/100 m for the region (Grover and Read, 1983). The lack of albite and presence of K-feldspar in samples of the Deicke and Millbrig from the Cincinnati Arch suggest that maximum temperatures experienced by the sediments in that region never reached much above 75° to 100°C, constraining the maximum depth of burial to about 2,100 m. Actual depth of burial was probably less, however, based on conodont CAI values as discussed above. Conversely, the abundance of authigenic albite along with K-feldspar in the K-bentonites throughout much of the southern Valley and Ridge implies that the low value for maximum temperature in that region was at least 70°C, and given the thicker Paleozoic sequence there as compared with the Cincinnati Arch, higher burial temperatures would be expected. The actual maximum temperature experienced by the K-bentonites and adjacent sediments in the Valley and Ridge cannot be determined from the results of the present study, but the presence of both authigenic K-feldspar and primary intermediate plagioclase in Deicke and Millbrig samples from the western Valley and Ridge suggests that there it was not much higher than 140°C. In the eastern Valley and Ridge, where the K-bentonites occur in clastics of the Blount Group, a maximum temperature cannot be determined based on feldspars because so few altered, unaltered, or authigenic feldspars are present. Therefore the findings of Harris et al. (1978) and Grover and Read (1983) provide the best approximation concerning the maximum temperatures experienced by the K-bentonites, probably between 200° and 300°C.

In Alabama and Georgia, diagenetic conditions may have remained favorable for authigenesis of K-feldspar into the Mississippian, because Upper Ordovician, Silurian, and Devonian strata are much thinner than their maximum thickness of 3,000 to 3,500 m farther north in Virginia. This difference in overburden thickness of the Middle Paleozoic sediments is indicated in Figure 22. It is in fact possible that greater burial depths were not reached in much of the southern Valley and Ridge until the late Mississippian or Pennsylvanian, when the thick sequence of coal-bearing strata was deposited along nearly the entire length of the Appalachian basin. Throughout this time burial depths along the Cincinnati Arch never exceeded approximately 1,200 m, and except in the Colvin Mountain Sandstone pore waters throughout the region were reducing.

During the Late Ordovician and Silurian, oxidizing meteoric waters moved into the eastern margin of the basin as river systems associated with the tectonic highlands to the east began to form a unified drainage system. These oxidizing waters flushed out the connate, reducing pore waters along much of the eastern part of the basin, a process that may have resulted in the oxidation of the pyrite and organic matter in the Moccasin and red Bays Formations. The ferroan minerals in the Deicke and Millbrig were affected by this change in redox potentials of the pore waters and, like the Moccasin and red Bays, the K-bentonites also became reddish colored as pyrite was oxidized to limonite, goethite and, finally, hematite. In the K-bentonites, sorption of this hematite occurred most commonly on the I/S. As the uplands were eroded, the influx of meteoric waters lessened, and the pore waters began to stagnate, but diagenesis of the ferroan and ferrotitanian minerals was little affected by these later changes. Throughout this time the pore waters continued to be reducing along the western Valley and Ridge and the Cincinnati Arch.

If it had not been forming already, authigenic albite probably began to form during the Mississippian or Pennsylvanian in the part of the Appalachian basin that is now the Valley and Ridge. As the overburden thickness and thus burial temperatures increased, and brines from the Sevier basin or elsewhere moved through the Ordovician sequence, the albite partly or completely replaced primary andesine or labradorite and existing authigenic K-feldspar (and primary K-feldspar, assuming that the ash originally contained it). Albitization began as temperatures in Ordovician sediments reached the threshold of albitization, between 75° and 125°C, when brines derived from dewatering of the Sevier basin moved westward toward the craton (Kesler et al., 1988). These brines probably catalyzed the formation of authigenic feldspars in the K-bentonites, and as they reacted with the smectite in the beds, the transformation to mixed-layer I/S in the K-bentonites also occurred (Elliott and Aronson, 1987). The illitization process was probably completed in a relatively short period (Morton, 1985), but the length of time needed for the various feldspathization reactions to go to completion is not known. Because the extent to which the illitization reaction goes to completion is a function of ambient burial temperature (Morton, 1985), it is evident that the highest temperatures at the time of illitization were along the southeastern edge of the Valley and Ridge, where the illite:smectite ratio in the Millbrig exceeds 9:1 in several samples from the Bays Formation. The lowest temperatures were along the Cincinnati Arch, where the illite:smectite ratio is between 3:1 and 4:1. During this entire time geochemical conditions along the Cincinnati Arch continued to be favorable only for the formation of authigenic K-feldspar, not albite.

In the Late Permian or Triassic, the tectonic setting changed as the regional compression abated, and an extensional tectonic regime began to affect the continental margin. Eventually, erosion began to exceed uplift and the younger sediments were stripped away, with some of this eroded material being redeposited in the sediments of the Newark Supergroup in the several Mesozoic basins of the Atlantic seaboard. As a result, the Ordovician strata and the associated

K-bentonites were exposed to increasingly shallower depths and lower temperatures, shown in Figure 22 by the upward trend of the burial paths. Although albite authigenesis would no longer have been occurring, it is possible that K-feldspar authigenesis began again if geochemical conditions were still favorable. More likely, however, the eventual passage through the basin of the potassic brines resulted in a change in the geochemistry of the pore waters, specifically a lowering of the $K^+/K^+ + H^+$ activity ratio to a level that no longer favored feldspar authigenesis (Kastner and Siever, 1979). As a result older authigenic albite and K-feldspar grains were "frozen" in the textural relationships seen at present, recording the geochemical conditions present during the time that maximum burial temperatures existed across the basin.

STRATIGRAPHY AND CORRELATION

Introduction

The discovery and discretionary use of marker beds is one of the principal methods of stratigraphy, for without them correlation of measured sections is very difficult indeed. A layer of airfall sediment such as volcanic ash or dust from a bolide impact is a nearly ideal marker bed because those sediments are usually dispersed over extremely large areas and no other depositional event can match the geologically instantaneous nature of their deposition. Problems can arise if several layers are deposited within a relatively short time span, an unlikely problem with meteoritic dust, but potentially a very real problem with volcanic ashfall deposits, as major eruptions can occur over many months or years and consist of several eruptive events, each of which may generate enough windborne ash to produce a recognizable marker bed (Fisher and Schmincke, 1984).

Although the preservation potential of individual ashfalls is not the same everywhere that the ash accumulates because of variations in local and regional sedimentation rates, wind or water currents active at the time of deposition, and postdepositional changes, such beds are nonetheless potentially excellent stratigraphic markers. Therefore, bentonites and K-bentonites like the Deicke and Millbrig, which are the alteration products of airfall volcanic ash, are ideal marker beds and potentially very useful in, for example, basin analyses, sedimentologic studies, and paleogeographic and paleotectonic reconstructions of stratigraphic sequences. The Deicke and Millbrig in particular have great potential for assisting such studies because of their tremendous lateral extent, which for the Millbrig is likely trans-Atlantic (Huff et al., 1992).

The stratigraphic framework for correlation locally in the Ordovician of the southeastern United States is principally lithostratigraphic, but the framework for correlation with Ordovician strata elsewhere in the country and the world is almost wholly biostratigraphic and is based principally on conodont and graptolite biozones. An ideal stratigraphic framework

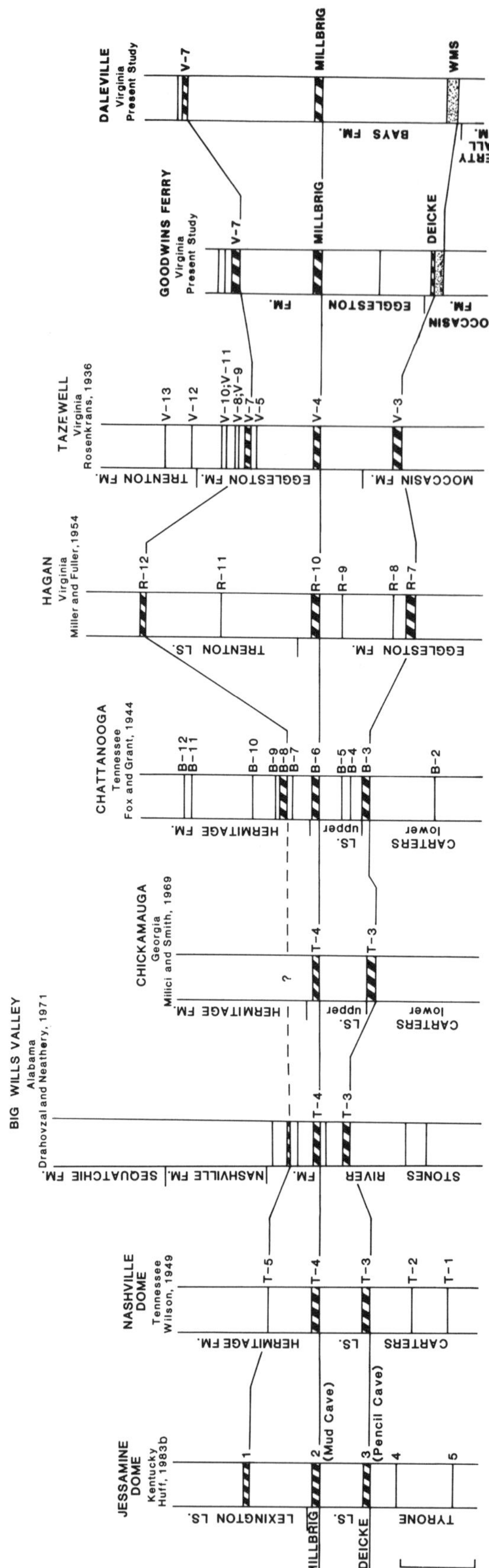

Figure 23. Summary correlation chart for several well-known measured sections in the southeastern United States in which Rocklandian K-bentonites occur. This chart is based on the results of the present study as shown in the following cross sections, and it represents a reinterpretation of Figure 8 of Huffman (1945), Plate 26 of Miller and Fuller (1954), and Plate 12 of Hergenroder (1966), and an expansion of Plate 10 of Rosenkrans (1936), Figure 2 of Fox and Grant (1944), Figure 7 of Huff and Kolata (1990), and Figure 22 of Haynes (1992). WMS-Walker Mountain Sandstone Member.

is, however, based on totally facies-independent isochrons such as individually recognizable beds of altered volcanic ash, and for such a framework to be accepted on more than a local scale, it is necessary to have a detailed understanding of some aspect of the petrology, geophysical character, etc., of each bed. Examples of how this can be done for the Rocklandian K-bentonites are Huff and Kolata (1990), who used geophysical logs of wells to demonstrate that the Deicke and Millbrig can be traced from the Upper Mississippi Valley to the Cincinnati Arch and the westernmost Valley and Ridge, and Haynes (1992 and present publication), who continued those correlations across the Valley and Ridge province on the basis of mineralogy and physical stratigraphy. Although there is some disagreement over the ability of chemical fingerprinting to identify individual beds of volcanic ash (altered and unaltered) accurately (Conkin et al., 1992a, b), the utility of single crystal geochemistry (Desborough et al., 1973; Yen and Goodwin, 1976; Samson et al., 1988), geophysical logs (Huff and Kolata, 1990), and chemical fingerprinting (Huff, 1983b; Cullen-Lollis and Huff, 1986; Kolata et al., 1986; Lyons et al., 1992a) as valid correlation tools in stratigraphic studies is widely accepted in geology.

One significant problem encountered in the stratigraphic literature is the uncertainty in some papers as to the identity of K-bentonite beds in exposures at stratigraphically equivalent but widely separated outcrops. For example several discrepancies in Plate 26 of Miller and Fuller (1954) have recently been addressed (Haynes, 1992), but even in separate publications by the same authors there can be significant differences in the identification of K-bentonites, as discussed in preceding text with respect to the Hounsfield "metabentonite" (Kay, 1931, 1935), and the K-bentonite sequence at the Big Ridge exposure near Duck Springs, Alabama (Drahovzal and Neathery, 1971; Neathery and Drahovzal, 1986). More recently, the lack of a clear understanding of regional stratigraphic relationships led to a likely misinterpretation of analytical results obtained from samples of the Deicke and Millbrig (Haynes and Huff, 1990).

The need for a synthesis of the regional stratigraphy of the Deicke and Millbrig to resolve such confusion is clear, and a discussion of the stratigraphic setting and new regional correlations of these two K-bentonite beds throughout the eastern midcontinent is presented in this section. The correlations are summarized in Figure 23, which shows the stratigraphic position of the Deicke and Millbrig in some of the best known exposures containing Rocklandian K-bentonites along the Cincinnati Arch and in the Valley and Ridge between Birmingham and Roanoke. The Deicke and Millbrig are identified in all the sections where present, and the sections use the Millbrig as a stratigraphic datum because it is the more widespread of the two beds. Figure 14 gives the legend for all cross sections. Figure 24 is an isopach map that accompanies a discussion of how variable the distribution of the Millbrig is in the central Kentucky outcrop.

Figures 25 through 33 show correlations of the Deicke and Millbrig K-bentonite Beds that are based on the study of over

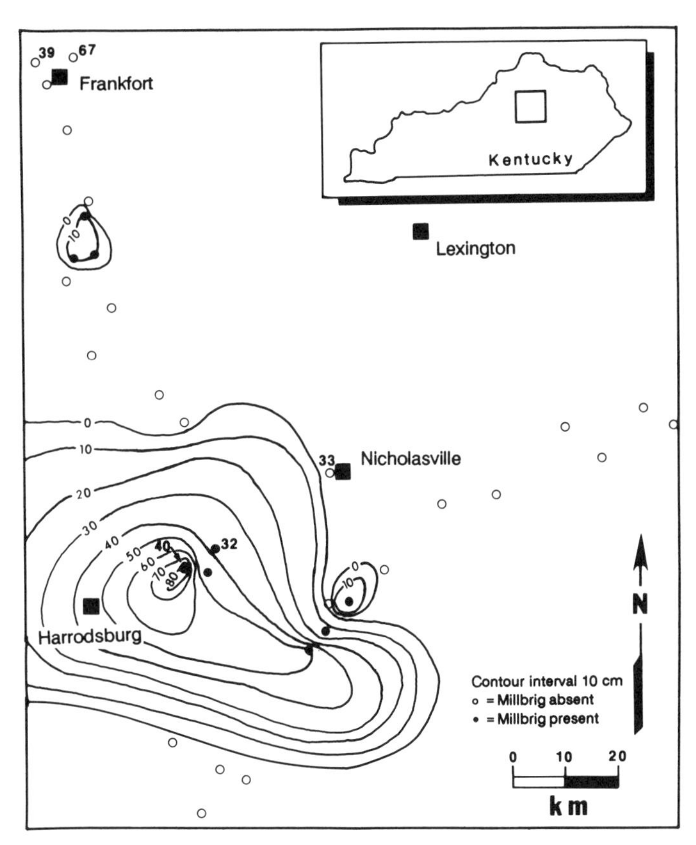

Figure 24. Isopach map of the Millbrig in the central Kentucky outcrop based on measurements of more than 30 exposures. The extreme local variation in thickness is most evident between the Shakertown (40), High Bridge (32), and Lexington Limestone Quarry (33) sections, as shown in Figure 25. Except in certain areas, such as here in central Kentucky, the thickness and lateral persistence of the Millbrig are very constant in the Cincinnati Arch and the southern Valley and Ridge. (Adapted and modified from Scharpf, 1990).

60 exposures of Rocklandian strata in the southeastern U.S. In several of these illustrations a unit identified as the Trenton ("Martinsburg") Formation is present. The term Martinsburg is used parenthetically because in southwestern Virginia and neighboring areas of West Virginia and Tennessee, the formal name Martinsburg is inappropriate (Diecchio, 1985; Walker and Diehl, 1985; Haynes, 1992). In this region the Martinsburg Formation of Butts (1940), Kreisa (1980), and other workers is the stratigraphic unit that conformably overlies the Eggleston or Moccasin Formation (western and central belts) and Bays Formation (eastern belt), and conformably underlies the Juniata or Sequatchie Formation, or unconformably underlies other younger units (Dennison et al., 1992). The lowest part of this unit is most commonly a sequence of limestones and lesser siltstones containing abundant Trenton-age fossils. This is in great contrast to the Martinsburg Formation in its type area of West Virginia and northern Virginia, where it is a deep water flysch sequence. Nonetheless, use of the name Martinsburg is unfortu-

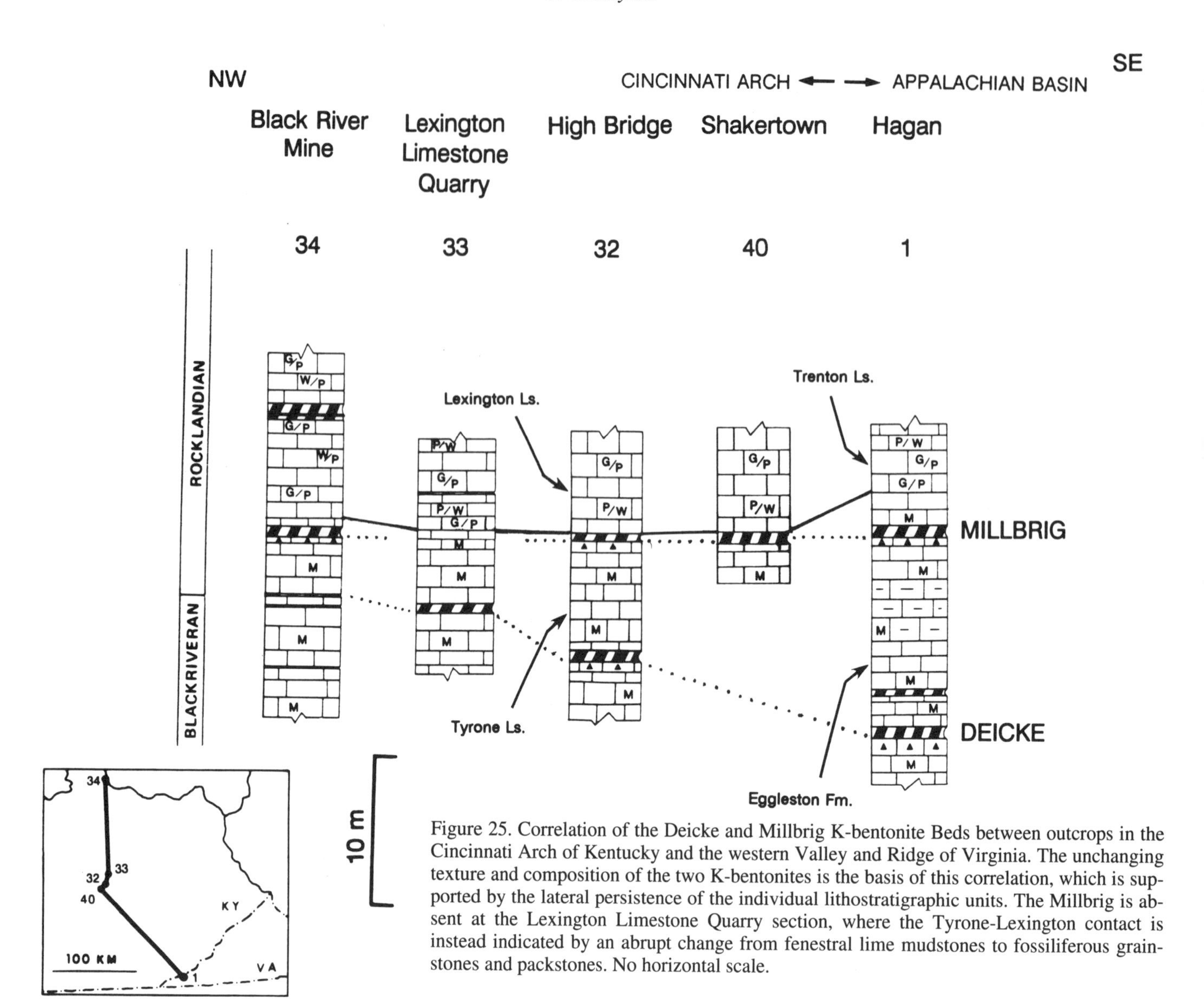

Figure 25. Correlation of the Deicke and Millbrig K-bentonite Beds between outcrops in the Cincinnati Arch of Kentucky and the western Valley and Ridge of Virginia. The unchanging texture and composition of the two K-bentonites is the basis of this correlation, which is supported by the lateral persistence of the individual lithostratigraphic units. The Millbrig is absent at the Lexington Limestone Quarry section, where the Tyrone-Lexington contact is instead indicated by an abrupt change from fenestral lime mudstones to fossiliferous grainstones and packstones. No horizontal scale.

nately widespread in southwestern Virginia and vicinity. The Trenton Formation is a better name for that unit than Martinsburg, but because Trenton is more appropriately a group name in much of this region (Perry, 1972), it is also a less than ideal name. Thus, the best nomenclatural choices for these strata are "Dolly Ridge Formation" (Perry, 1972) for the lower part, which is calcareous, contains Trenton-age fossils, and overlies the Eggleston/Moccasin/Bays interval, and "Reedsville Formation" (Reedsville Shale of Ulrich, 1911) for the upper part of the unit, which is more clastic and underlies the Juniata/Sequatchie interval. This significant stratigraphic problem needs a final resolution, because it continues to cause nomenclatural confusion on a regional scale.

It is intended that Figure 23 will join Figure 22 of Haynes (1992) in superceding many, if not all, older correlation charts showing suggested regional correlations of Rocklandian K-bentonite sequences in the southeastern U.S. Such charts include Plate 26 of Miller and Fuller (1954) and Plate 12 of Hergenroder (1966). The correlations shown in Figure 23 are based on the results of the present study, and the chart is intended to serve as a reference when referring to older publications. Thus, Figure 23 schematically shows the well-known "type" sections for the alpha-numeric designations previously applied to the Rocklandian K-bentonite sequence in the southeastern U.S. The names of the geologists who first (or best) described the K-bentonites at the particular exposure are given for reference, and the sections are correlated with each other based on the results of the present study.

Figures 34 and 35 are generalized chronostratigraphic correlation charts that show the relationship of the Deicke and

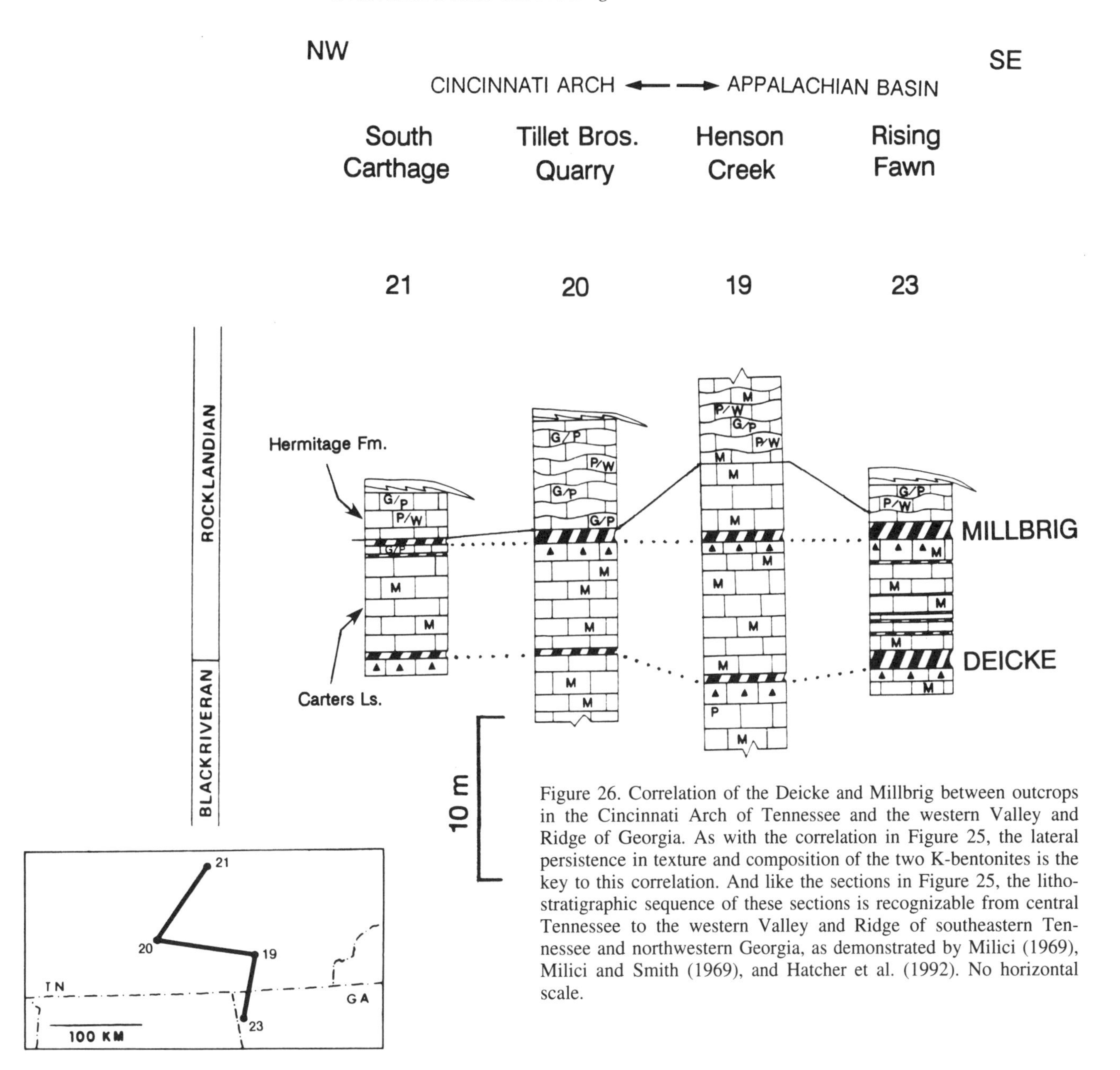

Figure 26. Correlation of the Deicke and Millbrig between outcrops in the Cincinnati Arch of Tennessee and the western Valley and Ridge of Georgia. As with the correlation in Figure 25, the lateral persistence in texture and composition of the two K-bentonites is the key to this correlation. And like the sections in Figure 25, the lithostratigraphic sequence of these sections is recognizable from central Tennessee to the western Valley and Ridge of southeastern Tennessee and northwestern Georgia, as demonstrated by Milici (1969), Milici and Smith (1969), and Hatcher et al. (1992). No horizontal scale.

Millbrig to Middle and Upper Ordovician strata in the eastern midcontinental U.S. Also included in the appropriate sections is the position of K-bentonite bed V-7 of Rosenkrans (1936) because it is a prominent K-bentonite and a useful marker in much of southwestern Virginia (Haynes, 1992). Figure 36 shows several local and regional unconformities, including three that are now better defined because of the recognition of the Deicke and Millbrig over much of the region. Figure 37 is a correlation chart from Hall et al. (1986) that is based on the conodont biostratigraphic work of Hall (1986). The chart has been modified herein to compare the biostratigraphic information with the lithostratigraphic data available from a consideration of K-bentonites, in this case, the Millbrig.

Absolute age of the Deicke and Millbrig

Based on ^{40}Ar/ ^{39}Ar values obtained from analyses of biotites in the Deicke and Millbrig, Kunk and Sutter (1984) determined that the mean isotopic age of the two beds is 454.1 ± 2.1 Ma. Samples used for their analyses included Deicke samples from exposures at South Carthage, Tennessee, and Big Ridge, Alabama, and Millbrig samples from the Black River Mine at Carntown, Kentucky, and Big Ridge, Alabama, exposures from which K-bentonites were sampled for the present study. And, using fission track dating of zircons in the beds, Ross et al. (1982b) determined a mean age of 453 ± 3 Ma for the stratigraphic interval in which the K-bentonites

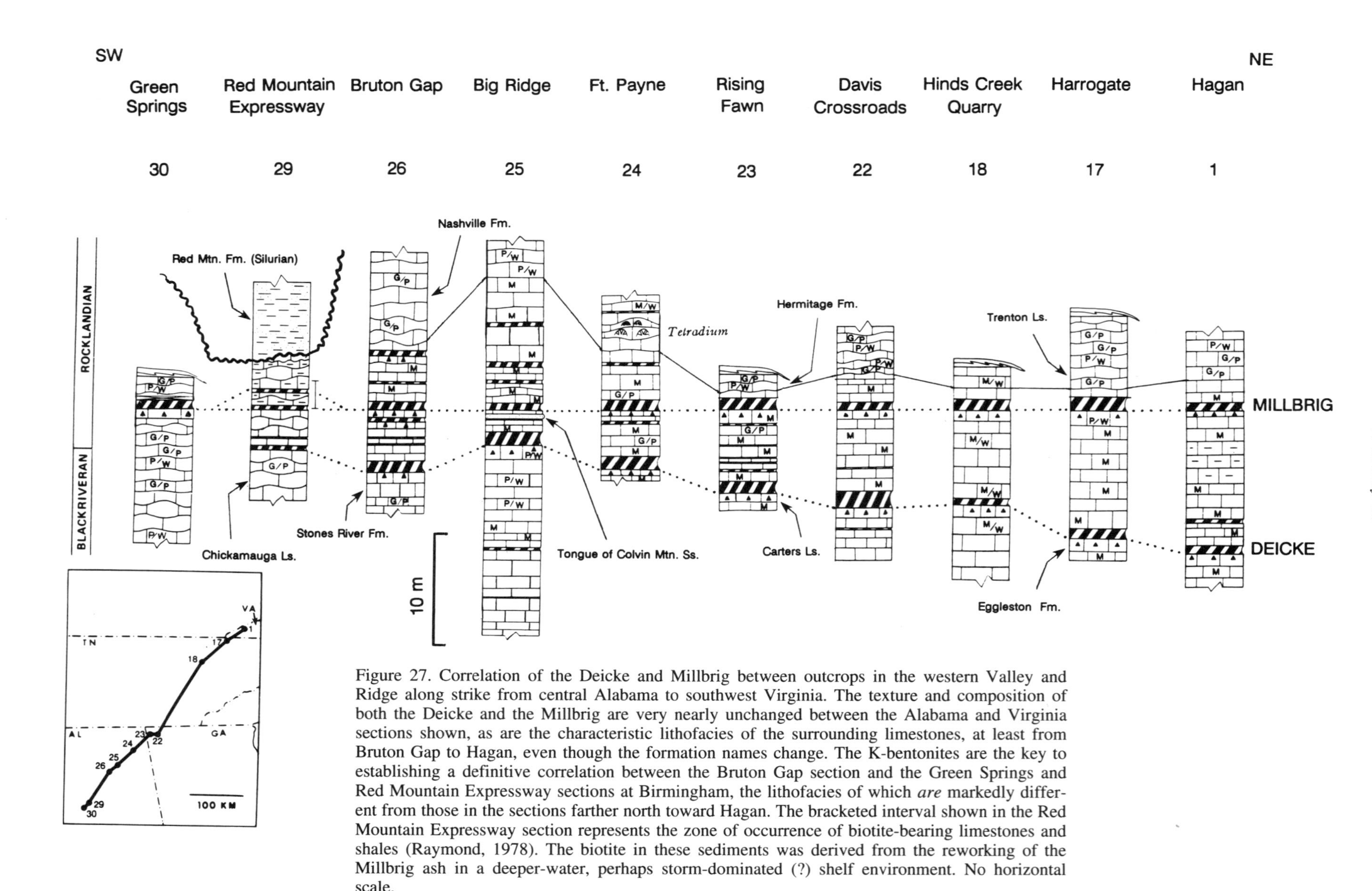

Figure 27. Correlation of the Deicke and Millbrig between outcrops in the western Valley and Ridge along strike from central Alabama to southwest Virginia. The texture and composition of both the Deicke and the Millbrig are very nearly unchanged between the Alabama and Virginia sections shown, as are the characteristic lithofacies of the surrounding limestones, at least from Bruton Gap to Hagan, even though the formation names change. The K-bentonites are the key to establishing a definitive correlation between the Bruton Gap section and the Green Springs and Red Mountain Expressway sections at Birmingham, the lithofacies of which *are* markedly different from those in the sections farther north toward Hagan. The bracketed interval shown in the Red Mountain Expressway section represents the zone of occurrence of biotite-bearing limestones and shales (Raymond, 1978). The biotite in these sediments was derived from the reworking of the Millbrig ash in a deeper-water, perhaps storm-dominated (?) shelf environment. No horizontal scale.

occur. Those two studies have constrained the numerical age of the Deicke and Millbrig to a very narrow range, and an age of 453 to 454 Ma clearly places the beds in the Caradocian of the British standard section, and the Rocklandian of the North American section.

The Rocklandian: Middle or Late Ordovician?

At their type sections in the Upper Mississippi Valley, the Deicke and Millbrig are in lower Rocklandian strata (Willman and Kolata, 1978), and in fact the Blackriveran-Rocklandian boundary in that area is considered by many geologists (Kolata et al., 1986; Sloan, 1987; 1988), but not all (Sweet, 1984), to be the base of the Deicke. Ross et al. (1982a) have shown that, based on the correlation of North American Ordovician sections with the standard European (British) section, the Rocklandian is correlative with part of the Caradocian, an Upper Ordovician stage. In North America, however, the Rocklandian is a stage of the Mohawkian (=Champlainian) Series, which has traditionally

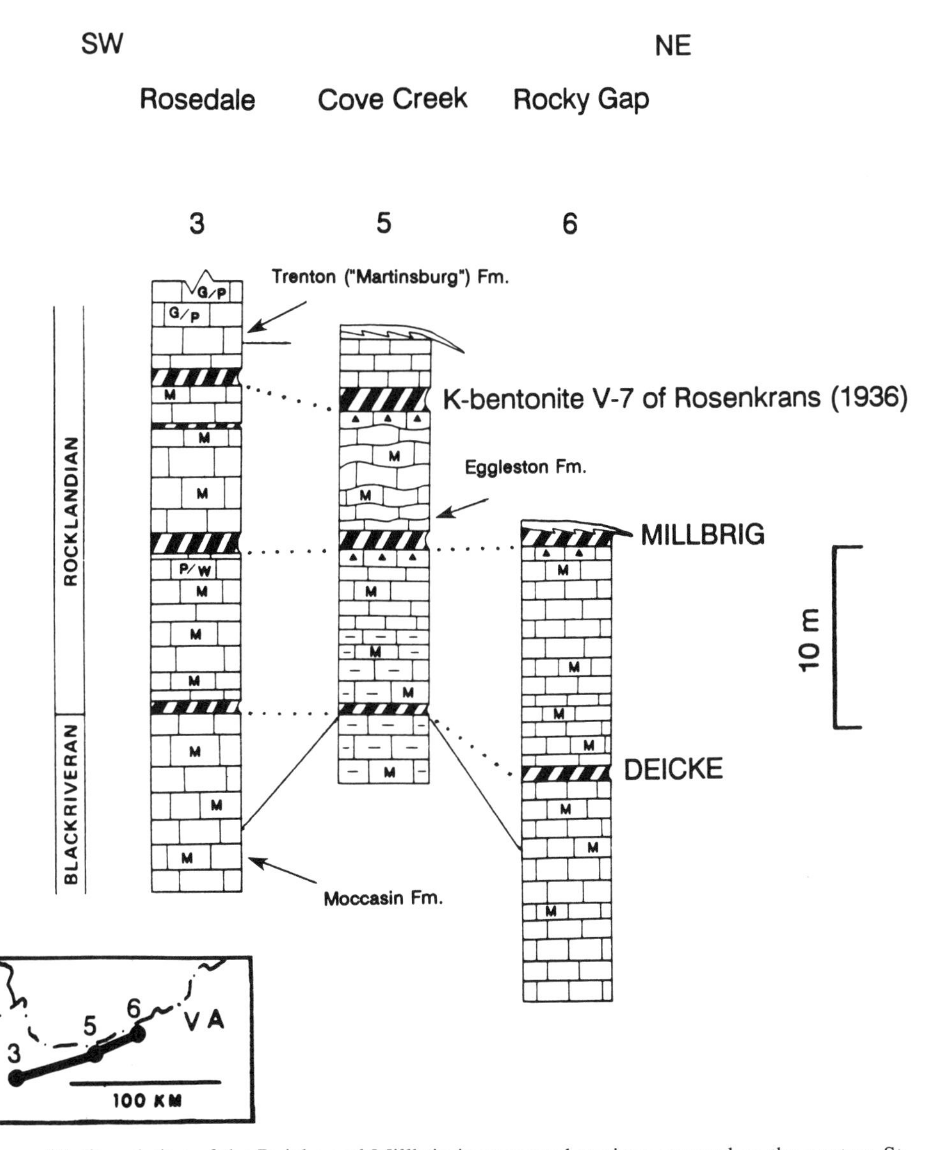

Figure 28. Correlation of the Deicke and Millbrig in measured sections exposed on the western St. Paul–Honaker thrust sheet (Rosedale) and the western Narrows thrust sheet (Cove Creek, Rocky Gap) in the central Valley and Ridge along strike in west-central Virginia. The stratigraphic sequence at these sections is transitional between the Powell Valley sequence at Hagan (Fig. 25), and the sections in the Clinch and New River Valleys farther east (Fig. 29). These sections and those in Figure 29 are in the area represented by the K-bentonite sequence shown in Plate 13 of Rosenkrans (1936). The section at Narrows, Virginia, on the St. Clair thrust sheet, which was described in detail by Rosenkrans (1936) but is now very poorly exposed and thus difficult to measure accurately, is also stratigraphically like the three sections shown here. No horizontal scale.

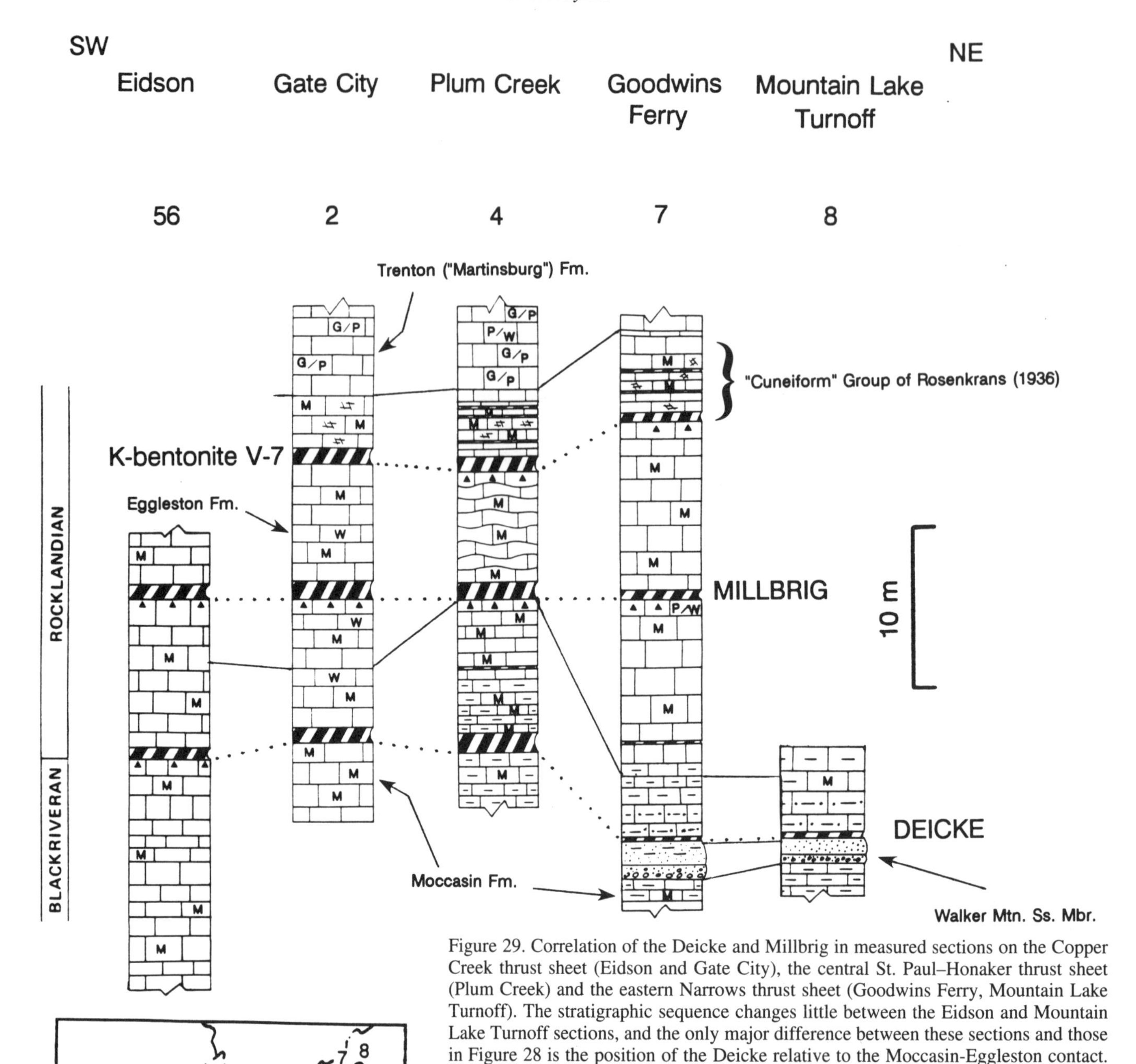

Figure 29. Correlation of the Deicke and Millbrig in measured sections on the Copper Creek thrust sheet (Eidson and Gate City), the central St. Paul–Honaker thrust sheet (Plum Creek) and the eastern Narrows thrust sheet (Goodwins Ferry, Mountain Lake Turnoff). The stratigraphic sequence changes little between the Eidson and Mountain Lake Turnoff sections, and the only major difference between these sections and those in Figure 28 is the position of the Deicke relative to the Moccasin-Eggleston contact. In the Rosedale, Cove Creek, and Rocky Gap sections in Figure 28, the Deicke is above that contact, whereas in sections to the east, shown here, the Deicke occurs below that contact, which becomes younger to the east and southeast. In the Narrows section the Deicke is also above the Moccasin-Eggleston contact (see Rosenkrans, 1936, Plate 13, and the position of bed V-3 therein). No horizontal scale.

been considered to be of Middle Ordovician age, although this standard is definitely changing, as a glance at the Ordovician System on any of the many *Correlation of Stratigraphic Units of North America* (COSUNA) charts will show, for example, Shaver (1985). So, I have considered Rocklandian strata to be Late Ordovician in age following this revised usage of Ross et al. (1982a) and the COSUNA charts in general, and Rader (1982) in particular for the Virginia Valley and Ridge. It is my hope that the proposed correlation of the Millbrig with the Big Bentonite of Scandinavia and the Baltic region (Huff et al., 1992) will be considered valid and will be acceptable to Ordovician geologists, thus unambiguously tying together the Mohawkian of eastern North America and the Caradoc of northern Europe and ending any debate over the age of Rocklandian strata in North America.

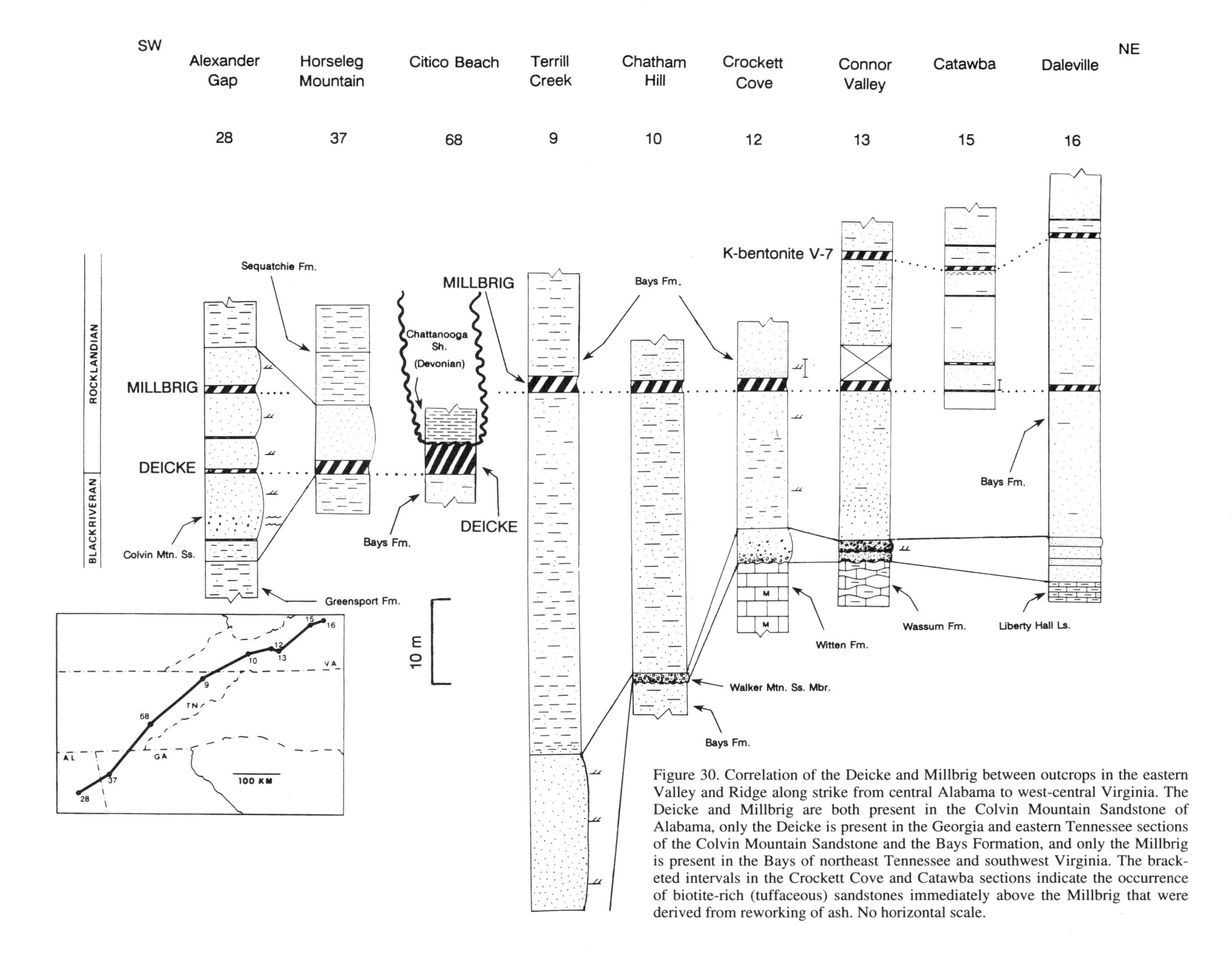

Figure 30. Correlation of the Deicke and Millbrig between outcrops in the eastern Valley and Ridge along strike from central Alabama to west-central Virginia. The Deicke and Millbrig are both present in the Colvin Mountain Sandstone of Alabama, only the Deicke is present in the Georgia and eastern Tennessee sections of the Colvin Mountain Sandstone and the Bays Formation, and only the Millbrig is present in the Bays of northeast Tennessee and southwest Virginia. The bracketed intervals in the Crockett Cove and Catawba sections indicate the occurrence of biotite-rich (tuffaceous) sandstones immediately above the Millbrig that were derived from reworking of ash. No horizontal scale.

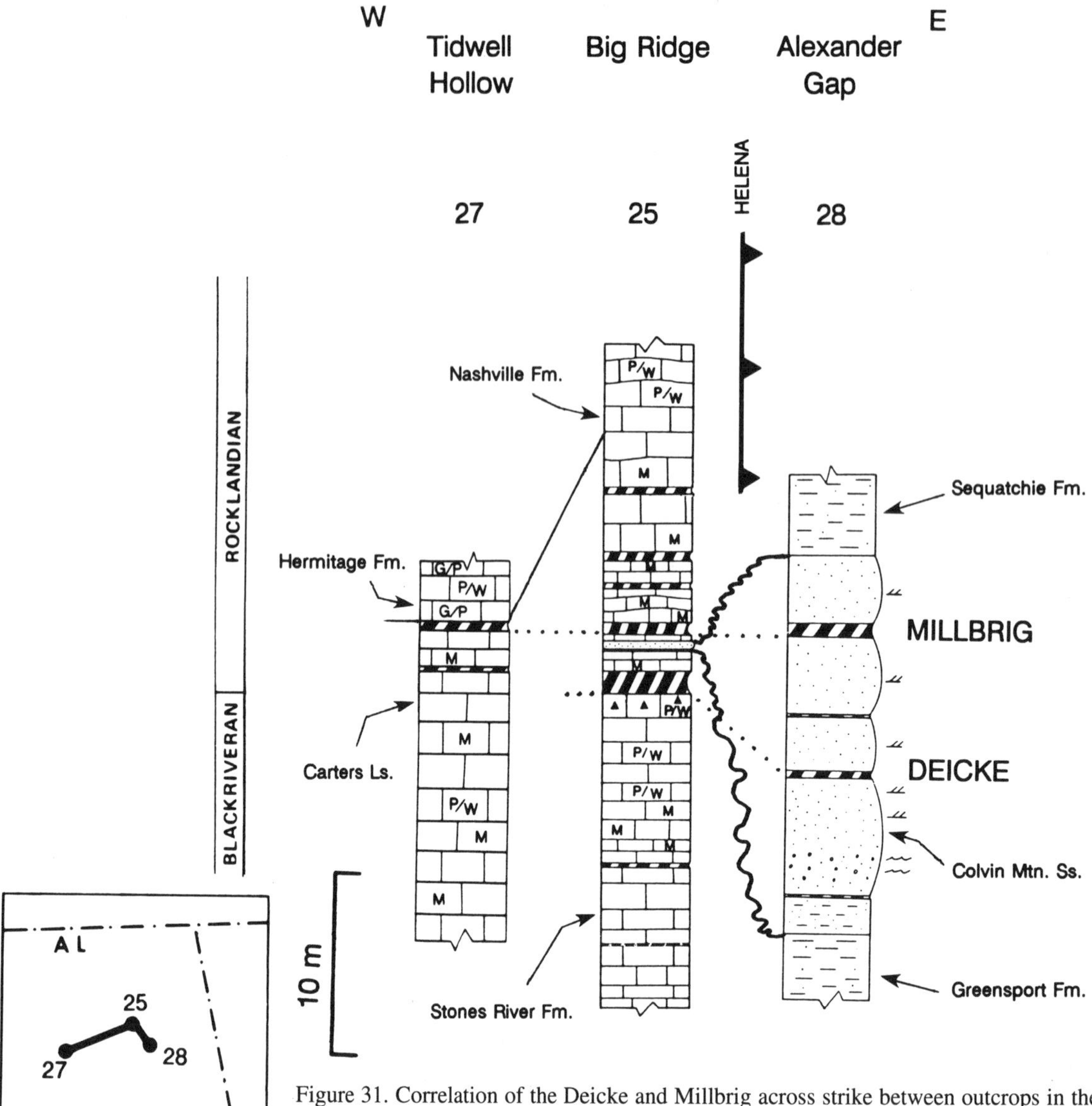

Figure 31. Correlation of the Deicke and Millbrig across strike between outcrops in the Valley and Ridge of Alabama. The Deicke is absent at Tidwell Hollow, and it is the Millbrig, with its characteristic composition, that is the key to the correlation of the Tidwell Hollow and Big Ridge sections. The presence of ilmenite and TiO_2 minerals in the Deicke is the key to correlation of the Big Ridge and Alexander Gap sections. This correlation, which is between sections on two of the major thrust sheets of the Alabama Appalachians and thus between carbonate and clastic sequences having little in common except the presence of K-bentonites, is reinforced by the occurrence of a 1- to 3-cm-thick tongue of the Colvin Mountain Sandstone between the Deicke and Millbrig in the Big Ridge section. The line of section crosses the Helena fault between the Big Ridge and Alexander Gap sections as shown. No horizontal scale.

Correlation along the Cincinnati Arch

Previous studies. Young (1940), McFarlan (1943), Kerr and Kulp (1949), Huff (1963, 1983b), Günal (1979), and Scharpf (1990) described the K-bentonites at several exposures in central Kentucky. McFarlan (1943) named the two thickest ones the "Pencil Cave bentonite" (older) and "Mud Cave bentonite" (younger), informal regional names that are still widely used by drillers and many others (Huff, 1983b). Wilson (1949) introduced the "T" system of nomenclature (T for Tennessee) for the K-bentonites in central Tennessee, and this nomenclatural system is currently used in eastern Tennessee and adjacent areas of Georgia and Alabama. (In the present study, petrographic examination of the Deicke and Millbrig K-bentonites began with samples collected from exposures along the Cincinnati Arch, and Wilson's work was particularly important because it provided the stratigraphic framework to use for my earliest attempts at correlating K-bentonites between the Cincinnati Arch and the western Valley and Ridge). Using trace element chemistry (chemical fingerprinting) and lithostratigraphic relations Günal (1979) and Huff (1983b) were able to correlate several K-bentonites collected from outcrops and cores along the northern Cincinnati Arch. The Deicke and Millbrig were correlated from the Upper Mississippi Valley to the western Valley and Ridge by Kolata et al. (1984) and Huff and Kolata (1990) on the basis of geophysical logs. The stratigraphic findings of these studies are supported by the findings presented herein.

Samson et al. (1988, 1989) suggested that the Deicke and Millbrig can be correlated on the basis of apatite chemistry, but concerns about the interpretation of some of their data, and its use for stratigraphic purposes, have been raised by Haynes and Huff (1990).

Discussion. In their type area in the Upper Mississippi Valley the Deicke and Millbrig are two of many Middle and Upper Ordovician K-bentonites, all of which are relatively thin (less than 10 cm). Of these, the Deicke and Millbrig are demonstrably the most laterally persistent (Kolata et al., 1986). Between the Upper Mississippi Valley and the Cincinnati Arch they become thicker and coarser grained, but their stratigraphic position is little changed (Huff and Kolata, 1990). In all regions the K-bentonites weather rapidly, forming noticeable reentrants in the outcrop, which are more noticeable with thicker beds. This recessive character of the K-bentonites is very noticeable even in relatively unweathered exposures (Figs. 2 through 5). In the subsurface, this tendency to recede or cave (the origin of the drillers' names Mud Cave and Pencil Cave) makes the K-bentonites recognizable on caliper logs as well as on the more common gamma ray and electric wireline logs (Huff and Kolata, 1990). Samples of the Deicke and Millbrig from cores and most outcrops along the Cincinnati Arch and in the western Valley and Ridge can be distinguished by their color, which for the Deicke ranges from dark green to grayish green (5G 5/2), and for the Millbrig from greenish gray to light olive gray (5Y 6/1). There are some variations in color between the coarse- and fine-grained zones of an individual bed at a given exposure, but the color differences between the Deicke and Millbrig tend to be quite distinctive regardless of grain size.

In the central Kentucky outcrop and subcrop of the Jessamine Dome the Deicke and Millbrig occur in the upper Tyrone Limestone. The Deicke is the only Rocklandian K-bentonite that is both laterally persistent and relatively thick (greater than 20 cm) in this area. It varies in thickness from just under 20 cm at the Black River Mine to just over 20 cm at exposures in the Frankfort area to over 70 cm in exposures along the main Kentucky River outcrop south of Lexington. By contrast, the distribution of the Millbrig is quite sporadic (Fig. 24) because of its position directly beneath the regionally extensive disconformity that is the Tyrone-Lexington contact. This contact, which is sharp and very distinct lithologically and is the defined position of the Millbrig in the central Kentucky outcrop (Scharpf, 1990), marks the horizon where the sparsely fossiliferous, fenestral, massively bedded micrites of the Tyrone are succeeded abruptly by the abundantly fossiliferous grainstones, packstones, and wackestones of the thinner bedded Lexington.

The isopach of the Millbrig illustrates the great lateral variability in the bed's thickness over short distances in the central Kentucky outcrop (Fig. 24). The greatest measured thickness regionally is 83 cm at the Shakertown section, yet at the High Bridge section only 3 km to the northeast, the Millbrig is 38 cm thick and, at the Lexington Limestone Quarry section another 12 km to the northeast, the Millbrig is completely absent (Figs. 24 and 25). The Millbrig then reappears in the subsurface farther north, where it reaches thicknesses of over 70 cm in cores, and in the subsurface exposure at the Black River Mine along the Ohio River (Fig. 25), where it is 73 cm thick. It is absent at exposures in the Frankfort area (Cressman and Noger, 1976; Scharpf, 1990), including the Taylor Avenue Quarry and Cove Spring Road sections (Appendix 1). This great lateral variability is unusual for a bed whose thickness and lateral continuity are otherwise quite constant over hundreds of square kilometers (Haynes, 1992) and is a testament to the erosion that occurred subsequent to deposition of the ash (Scharpf, 1990).

The erosional nature of the Tyrone-Lexington contact is also indicated by other features. At sections where the Millbrig is absent, the basal beds of the overlying Lexington contain rounded clasts of reworked Tyrone; at exposures where the Millbrig is present, such fragments are not observed (Scharpf, 1990). Also, variations in the thickness of the stratigraphic interval between the base of the Deicke and the Tyrone-Lexington contact suggest an erosional contact. Where the Millbrig is present, this interval is consistently about 7 to 8 m thick, regardless of the thickness of the Millbrig, but at exposures where the Millbrig is absent, the thickness is as little as 5.5 m (Cressman, 1973). This suggests that up to 3 m of the uppermost Tyrone were removed prior to deposition of the Lexington, and that the intensity of the erosion varied greatly over relatively short distances.

The evidence also indicates that erosion occurred subsequent to deposition of the full thickness of Millbrig ash. At exposures where the Millbrig is present but less than about 15 cm thick, it is always the middle and upper Millbrig, with its distinct and readily recognizable internal stratigraphy, that is missing (Scharpf, 1990). At such exposures the only recognizable diagnostic feature may be very thin remnants of the zone of coarse, biotite-rich tuffaceous material that characterizes the middle Millbrig throughout the study area (Figs. 15 and 19).

The central Kentucky lithostratigraphy is persistent throughout much of the midcontinent region, even though formation names change (Wilson, 1949, 1991; Gutstadt, 1958). This makes correlation relatively easy once one is familiar with the lithostratigraphy. In the Nashville Dome of Tennessee the Deicke and Millbrig occur in the Carters Limestone, a fen-

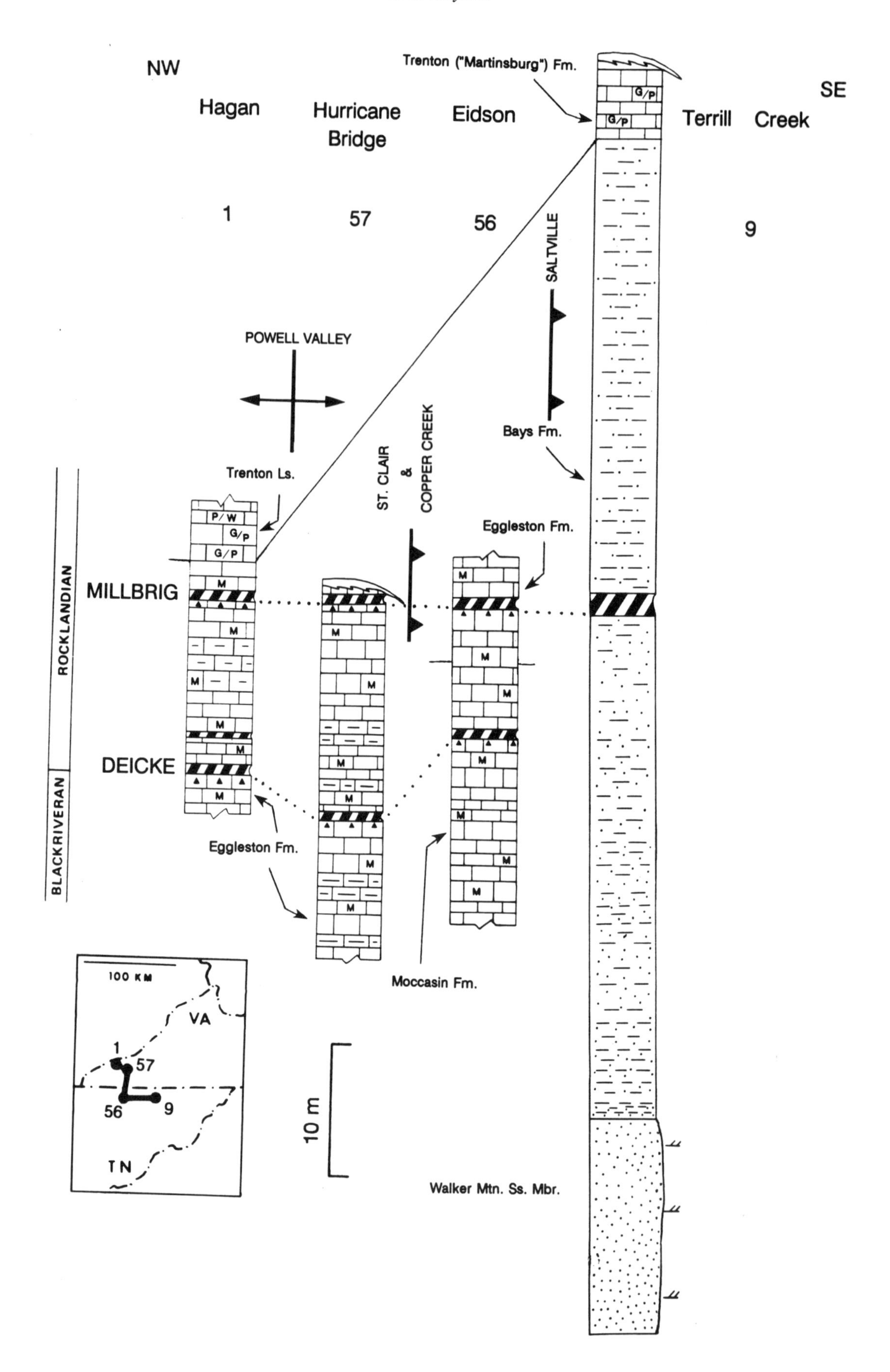

NW
SE
Hagan
Hurricane Bridge
Eidson
Terrill Creek
1
57
56
9
Trenton ("Martinsburg") Fm.
SALTVILLE
POWELL VALLEY
ST. CLAIR & COPPER CREEK
Bays Fm.
Trenton Ls.
Eggleston Fm.
ROCKLANDIAN
BLACKRIVERAN
MILLBRIG
DEICKE
Eggleston Fm.
Moccasin Fm.
100 KM
VA
TN
10 m
Walker Mtn. Ss. Mbr.

estral lime mudstone equivalent to the Tyrone of Kentucky. The Deicke occurs at the boundary between the lower and upper Carters. Where present, the Millbrig occurs at or just beneath the disconformable contact between the Carters and the overlying Hermitage Formation, the equivalent of the Lexington Limestone. At many exposures in central Tennessee, as in central Kentucky, the Millbrig commonly is thinned greatly or is even absent because of its occurrence beneath the regionally widespread disconformity at the base of the Hermitage/Lexington lithofacies (Wilson, 1949, 1991). And, just as in central Kentucky, there are rounded clasts of Carters in the basal Hermitage at exposures where the Millbrig and part of the underlying Carters Limestone have been removed by erosion (Wilson, 1991). No study has been done to determine how much of the upper Carters is missing in those sections where the Millbrig is absent, but in and around Nashville only 3 m or so of upper Carters is present between the Deicke and the base of the Hermitage (Wilson, 1991), suggesting that erosion in central Tennessee was appreciably greater than in central Kentucky. Because of this erosion, most occurrences of the Millbrig in the Nashville Dome are found along the eastern margin of the outcrop belt, and the thicknesses of the Deicke and Millbrig tend to be similar to those measured in the western Valley and Ridge. Some lateral variation does occur: in the highwall exposure at Shelbyville (Fig. 2B) the Millbrig is almost 80 cm thick, but the Deicke is only about 30 cm thick.

This stratigraphy persists into the Sequatchie Valley and the Valley and Ridge of Alabama and Georgia (Milici, 1969; Milici and Smith, 1969; Drahovzal and Neathery, 1971). In these areas both beds still occur in the Carters (or the readily recognizable Carters-like facies of the Stones River Group undivided). The Deicke occurs at the contact between the lower and upper Carters, and the Millbrig occurs at or just beneath the contact between the upper Carters and the overlying Hermitage. With few exceptions both beds tend to be consistently thick (over 50 cm) in this region, unlike the exposures along the Cincinnati Arch. The only significant lithostratigraphic change is that 1 to 2 m of Carters is present above the Millbrig at some exposures, for example, the Davis Crossroads section in Georgia (Fig. 27). This is because the Carters-Hermitage contact is gradational in much of this region, although in some of the westernmost exposures, for example, at the Tidwell Hollow section in Alabama, the contact is as abrupt as it is along the Cincinnati Arch, and the Millbrig immediately underlies the Hermitage (Fig. 2C).

Figure 32. Correlation of the Deicke and Millbrig across strike between outcrops in the Valley and Ridge of southwest Virginia and northeast Tennessee. There is little change in the stratigraphy between the Hagan section, on the western limb of the Powell Valley anticline, and the Hurricane Bridge section, on the eastern limb, but the changes between that section, the Eidson section, and the Terrill Creek section are more significant. In all sections, the presence of the Millbrig, with its characteristic biotite, is the key to this correlation, which is between sections on three of the major thrust sheets of the southern Appalachians. Also note that the base of the Trenton Formation is a useful marker horizon because of its lateral persistence across strike in this area. The line of section crosses the axis of the Powell Valley anticline between the Hagan and Hurricane Bridge sections, the St. Clair and Copper Creek faults between the Hurricane Bridge and Eidson sections, and the Saltville fault between the Eidson and Terrill Creek sections as shown. No horizontal scale.

Correlation from the Cincinnati Arch to the Valley and Ridge

Previous studies. Rosenkrans (1936) introduced the "V" system of nomenclature (V for Virginia) for K-bentonites in southwest Virginia, southeast West Virginia, and northeast Tennessee. He correlated several sections based on detailed study of the physical stratigraphy of the K-bentonite sequences (Rosenkrans, 1936, Plate 13). The beds he termed V-3, V-4, and V-7 are relatively thick and laterally persistent in the central belt, although across strike in the Powell Valley (western belt) beds V-1 through V-6 were thought by him to be absent because of a major erosional unconformity. Fox and Grant (1944) studied Ordovician strata around Chattanooga, Tennessee. They introduced the "B" nomenclatural system (B for bentonite), and the beds they termed B-3, B-6, and B-8 are the thickest and most widespread in that region. Of primary interest to the present study are their descriptions of those beds and their Figure 2, in which several correlations of Middle Ordovician strata from southeast Tennessee to surrounding areas were based on the K-bentonite sequence. A stratigraphic column based on that figure is shown in Figure 23. Huffman (1945) described the Ordovician limestones of the Powell Valley in Lee County, Virginia, in detail, and included the section at Harrogate just across the Tennessee state line. Those rocks were then correlated with equivalent strata described from the Jessamine Dome of central Kentucky. Huffman's study is particularly important because it was the first to describe in detail the section at Hagan, Virginia, with its well-exposed K-bentonite sequence. Miller and Fuller (1954) and Miller and Brosgé (1954) studied the geology of Lee County, Virginia, introducing yet another nomenclatural system for the K-bentonites, the "R" system (R for Rose Hill). The thickest K-bentonites in that area are beds R-7, R-10, and R-12. Miller and Fuller and Miller and Brosgé, like Huffman (1945), described the well-exposed section at Hagan in detail. In addition, they presented the first description of the generalized K-bentonite sequence in the Powell Valley, and Plate 26 of Miller and Fuller was the first attempt at unifying the various K-bentonite nomenclatural schemes by correlating K-bentonite sequences from several well-studied sections in the southeastern United States. Milici (1969) and Milici and Smith (1969) studied the Chickamauga Supergroup in its type area near Chattanooga and in the nearby Sequatchie Valley of

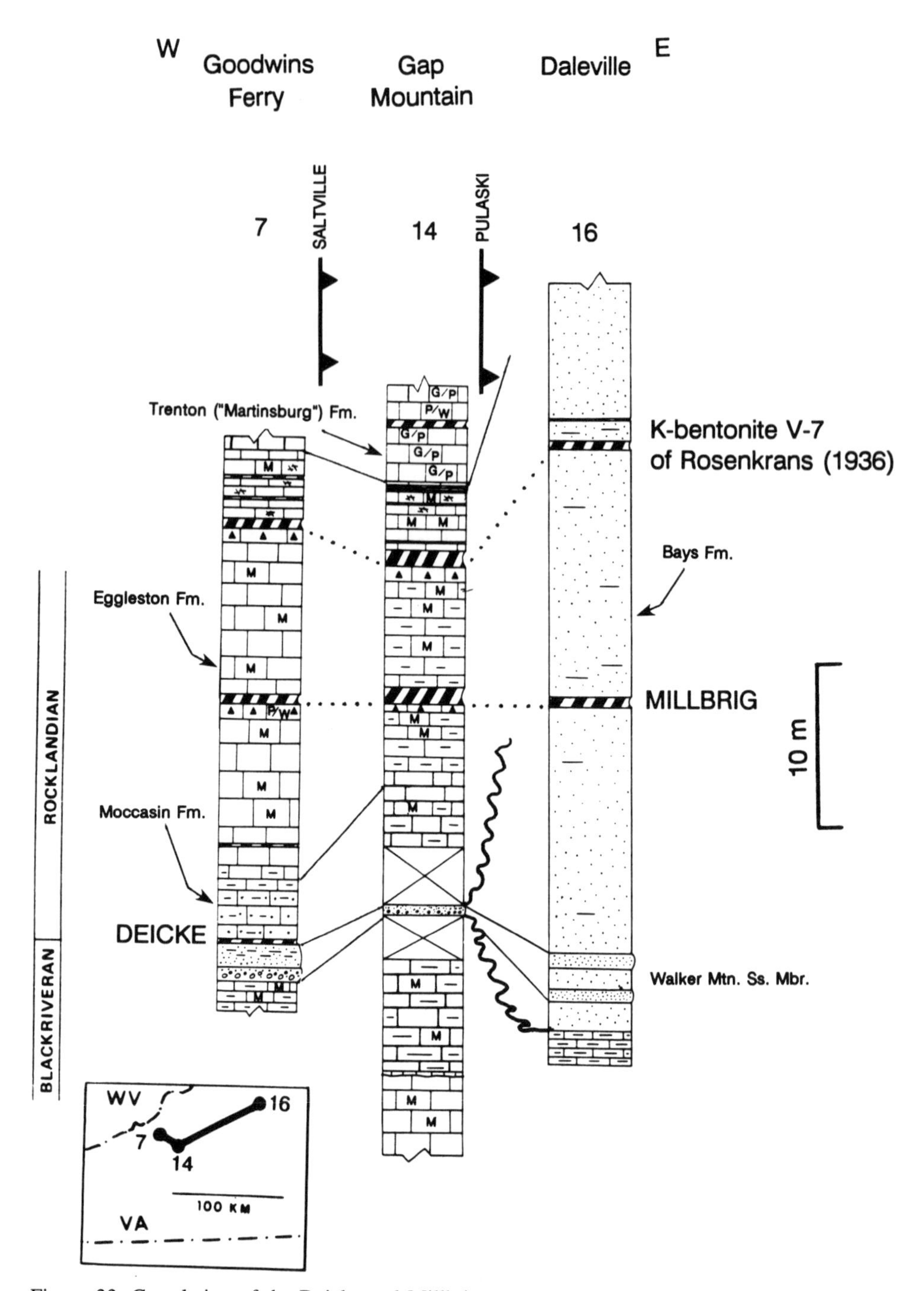

Figure 33. Correlation of the Deicke and Millbrig across strike between outcrops in the Valley and Ridge of west-central Virginia. Because the Deicke is very fine grained in this region, and can be easily mistaken for a red shale, it is the distinct mineralogy of the Millbrig, readily recognizable at most exposures, that is the key to this correlation. The Walker Mountain Sandstone Member is also a very useful marker in this region because, like the K-bentonites, it can be identified on four of the major thrust sheets of the Virginia Appalachians. The line of section crosses the Saltville fault between the Goodwins Ferry and Gap Mountain sections, and the Catawba fault, a branch of the Pulaski thrust system, between the Gap Mountain and Daleville sections as shown. No horizontal scale.

Tennessee. They used the Central Basin T series K-bentonite terminology of Wilson (1949) rather than the B series of Fox and Grant (1944), which was developed in the Chattanooga area, because the formation names from the Central Basin were also used. The K-bentonite beds T-3 and T-4 of Wilson were recognized and correlated from the Nashville area to the Sequatchie Valley and beyond to Chattanooga and northwest Georgia. Abundant biotite in bed T-4 at those sections was noted. Like Milici and Smith (1969), the guidebook by Drahovzal and Neathery (1971) used the K-bentonite terminology of Wilson (1949) rather than that of Fox and Grant (1944) for the K-bentonites in the Valley and Ridge of Alabama. Several of the sections in Figures 27, 30, and 31 are discussed in that guidebook. Most subsequent publications have also referred to the two most prominent K-bentonites in the Ordovician of Alabama and Georgia as T-3 and T-4 (Chowns and McKinney, 1980; Chowns and Carter, 1983a; Benson and Stock, 1986).

Although Huff (1983b) studied K-bentonite samples only from along the Cincinnati Arch, his Table 1 nonetheless shows suggested correlations of the Cincinnati Arch K-bentonites with those described from exposures elsewhere in the region, including those in the sequences of Fox and Grant (1944), Huffman (1945), and Wilson (1949). That table sums up what has been for decades the most widely accepted correlations of the K-bentonites throughout the region.

Discussion. Few studies have presented detailed lithologic, stratigraphic, or petrographic evidence that supports the correlation of K-bentonites between the Cincinnati Arch and the Valley and Ridge. As discussed by Milici and Smith (1969) for Tennessee, the Rocklandian sequence around Chattanooga is lithologically very similar to the Central Basin sequence, and so correlation between those areas is readily demonstrated. Elsewhere in the southern Valley and Ridge province, however, correlations are not made as easily. More than two thick K-bentonites occur, strata are not flat-lying as they are along the Cincinnati Arch, lithologies change markedly between outcrops, especially those on different thrust sheets, and most exposures are structurally disrupted to varying degrees. Color is of little use, because in some exposures the K-bentonites are red, in some they are yellowish tan to orange to brownish gray, and in still others they are a very pale green or grayish white. Furthermore, K-bentonites on one thrust sheet commonly are a different color than those on an adjacent sheet.

Because of these and other stratigraphic and structural difficulties, all of which have been noted to varying extents in previous studies, the Hagan section in the Powell Valley, in the western belt of Virginia, has proved to be the most important exposure for demonstrating regional correlations of the Deicke and Millbrig (Haynes, 1992). First, the Rocklandian strata of the western belt are lithologically and faunally transitional between sections farther east in the Valley and Ridge, and ones farther west along the Cincinnati Arch. Of the expo-

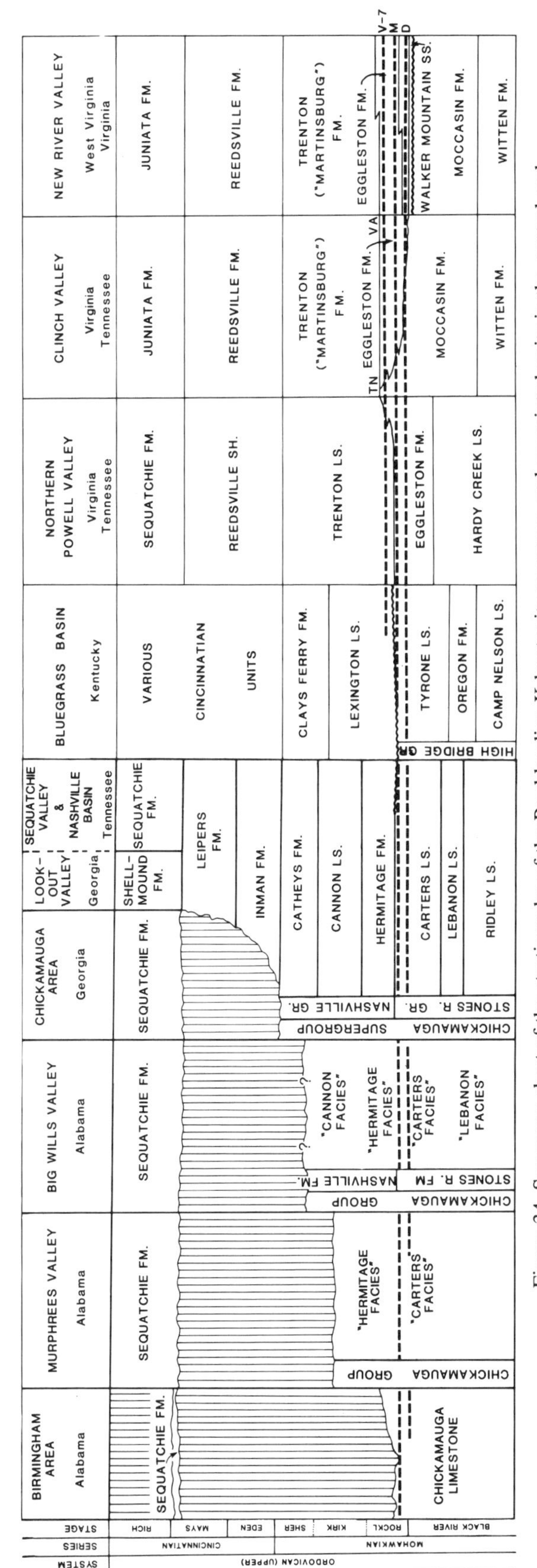

Figure 34. Summary chart of the stratigraphy of the Rocklandian K-bentonite sequence and associated units in the central and western belts of the southern Valley and Ridge province, and the Cincinnati Arch. To the right of the chart: V-7 = K-bentonite V-7 of Rosenkrans (1936), M = Millbrig, D = Deicke. Carbonate strata predominate in all sections in these areas.

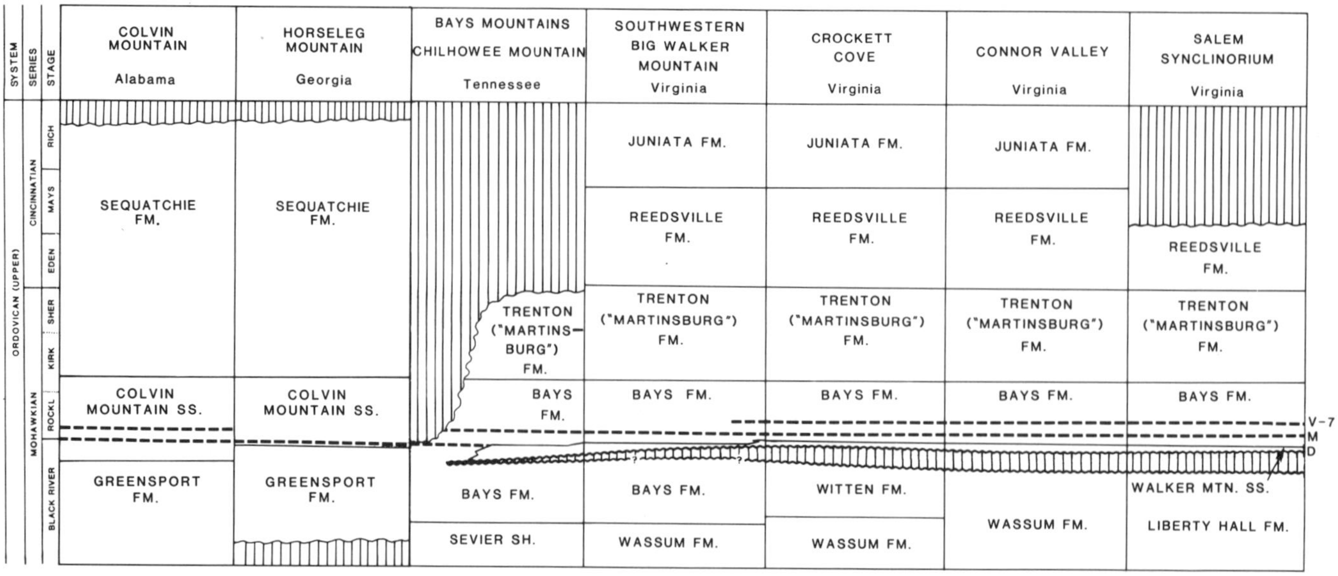

Figure 35. Summary chart of the stratigraphy of the Rocklandian K-bentonite sequence and associated units in the eastern belt of the southern Valley and Ridge province. To the right of the chart: V-7 = K-bentonite V-7, M = Millbrig, D = Deicke. Clastic strata of the Blount Group predominate in all sections. The Millbrig is the most laterally persistent K-bentonite in this belt, with K-bentonite bed V-7 of Rosenkrans (1936) being restricted to Virginia, and the Deicke being present only in the Colvin Mountain Sandstone of Alabama and Georgia and the Bays Formation of the Chilhowee Mountain area (Tellico-Sevier belt of Neuman, 1955) in east Tennessee.

sures in the Powell Valley, the one at Hagan is unquestionably one of the best and there is no doubt as to where in the section the Deicke and Millbrig are located (Fig. 5A), as demonstrated by the agreement of measured sections reported in the publications noted above. Second, samples of the Deicke and Millbrig from the Hagan section are more similar texturally to samples of the Deicke and Millbrig from central Kentucky, over 150 km to the west, than they are to samples of the K-bentonites from the central belt of the Valley and Ridge, just 20 to 25 km to the east. This remarkable similarity is mirrored to a great extent by similarities in the surrounding strata themselves, an indication of the great lateral persistence of facies between the westernmost Valley and Ridge and the Cincinnati Arch. Third, samples of the Deicke and Millbrig from the western outcrop belts of the Valley and Ridge in Alabama and Georgia also resemble samples from the Cincinnati Arch and thus are readily matched with samples from Hagan. Finally, because there are evidently no Rocklandian strata in the eastern belt of Alabama and Georgia (Hall et al., 1986), it is the Hagan section (Figs. 25, 27, and 32), rather than, for example, the also very well exposed sections at Big Ridge, Alabama (Figs. 27 and 31) or at Davis Crossroads, Georgia (Figs. 26 and 27), that is the key to extending the correlation of the K-bentonites from the shelf carbonate sequence of the Cincinnati Arch and the western Valley and Ridge across one or more thrust sheets into the redbed sequence of the central and eastern Valley and Ridge.

The Tyrone-to-Lexington sequence discussed above is readily recognizable at Hagan as the Eggleston-to-Trenton sequence because, as first noted by Huffman (1945), there is only slight change in the lithostratigraphy of this stratigraphic interval between central Kentucky and the Powell Valley. The major difference is that the Eggleston-Trenton contact is gradational over a 1- to 2-m interval, whereas the Tyrone-Lexington contact is erosional and therefore abrupt. At Hagan, three thick K-bentonites are present as shown in Figure 23; two in the Eggleston about 8 m apart (beds R-7 and R-10 of Miller and Fuller, 1954), and one in the Trenton (bed R-12 of Miller and Fuller), about 20 m above the upper bed in the Eggleston. The upper of these three thick beds is at present very poorly exposed and no samples could be obtained, so that bed is omitted from Figures 25, 27, and 32. Based on the petrographic criteria in Table 1, the lower bed, R-7, is the Deicke and the middle bed, R-10, is the Millbrig. The upper bed, R-12, was reported to contain some biotite by Miller and Fuller (1954), and as shown in Figure 23 it is correlated eastward with K-bentonite V-7 of Rosenkrans (1936). The correlation between Hagan and central Kentucky (Fig. 25) is demonstrated with relative ease considering the distance between Hagan and the exposures along the Kentucky River (Haynes, 1992).

Similarly, correlation of the Nashville Dome sections with the Sequatchie Valley and Rising Fawn sections in the Valley and Ridge can also be demonstrated relatively easily, considering the intervening distance (Fig. 26). The lithostratigraphy of the Rising Fawn section is very similar to equivalent sections of the Nashville Dome, but an even more similar sequence is at

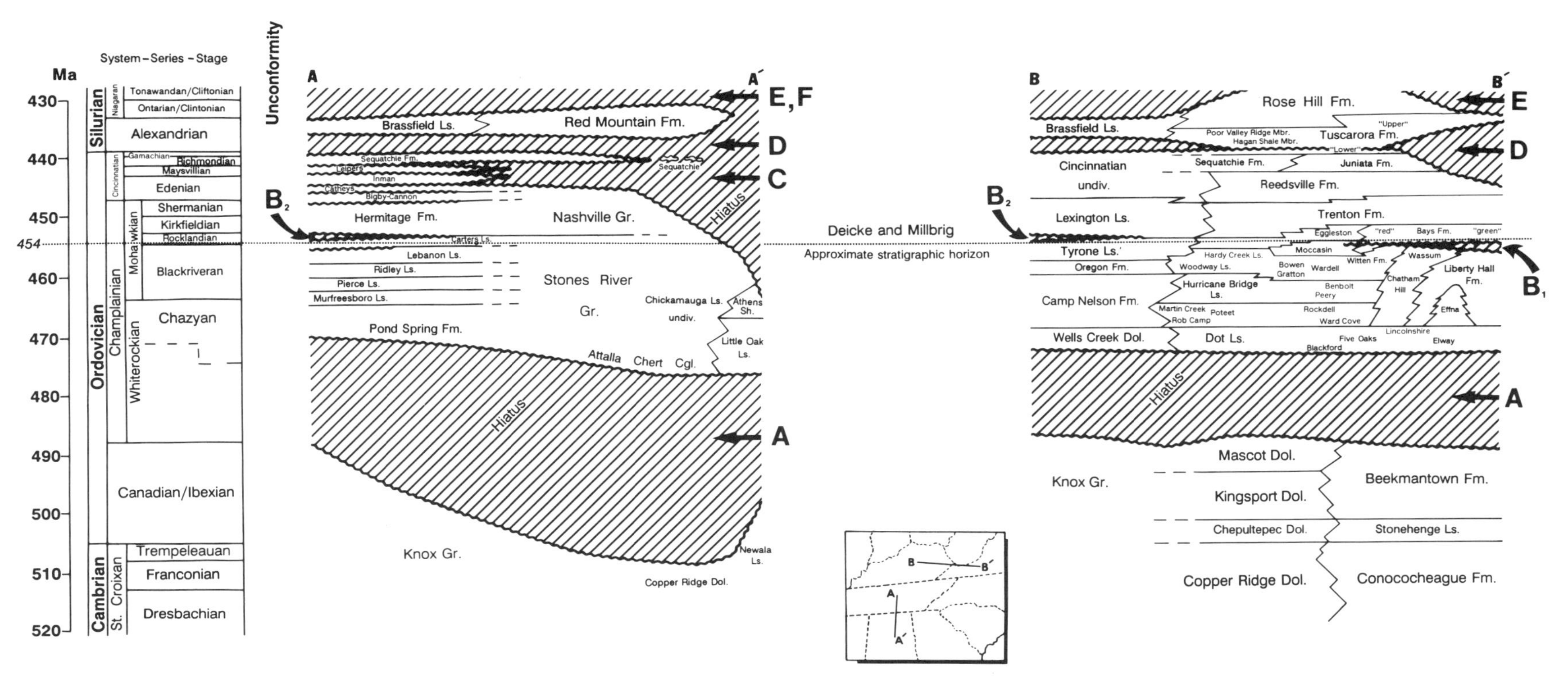

Figure 36. Stratigraphic relationships in the eastern midcontinent of Upper Cambrian through Lower Silurian strata. The approximate stratigraphic position of the Deicke and Millbrig K-bentonites and of recognized unconformities in the section are shown in both sections. Section A-A′, The major regional unconformities are the post-Knox, sub-Silurian, sub-Devonian, and sub-Mississippian, labeled A, D, E, and F, respectively; the Deicke and Millbrig are associated with two lesser unconformities, the post-Tyrone (labeled B_2) and the post-Chickamauga (labeled C). Section B-B′, The major unconformities are the post-Knox, sub-Silurian, and sub-Devonian, the same in Tennessee and Alabama, again labeled A, D, and E, respectively. As they are farther south, the Deicke and Millbrig are associated with two lesser unconformities, but only one, the post-Tyrone unconformity, labeled B_2, is present in both regions. There is no equivalent to unconformity C (post-Chickamauga) of Alabama, but in much of southwestern Virginia a sub-Walker Mountain Sandstone Member unconformity (labeled B_1) occurs downsection from the K-bentonites.

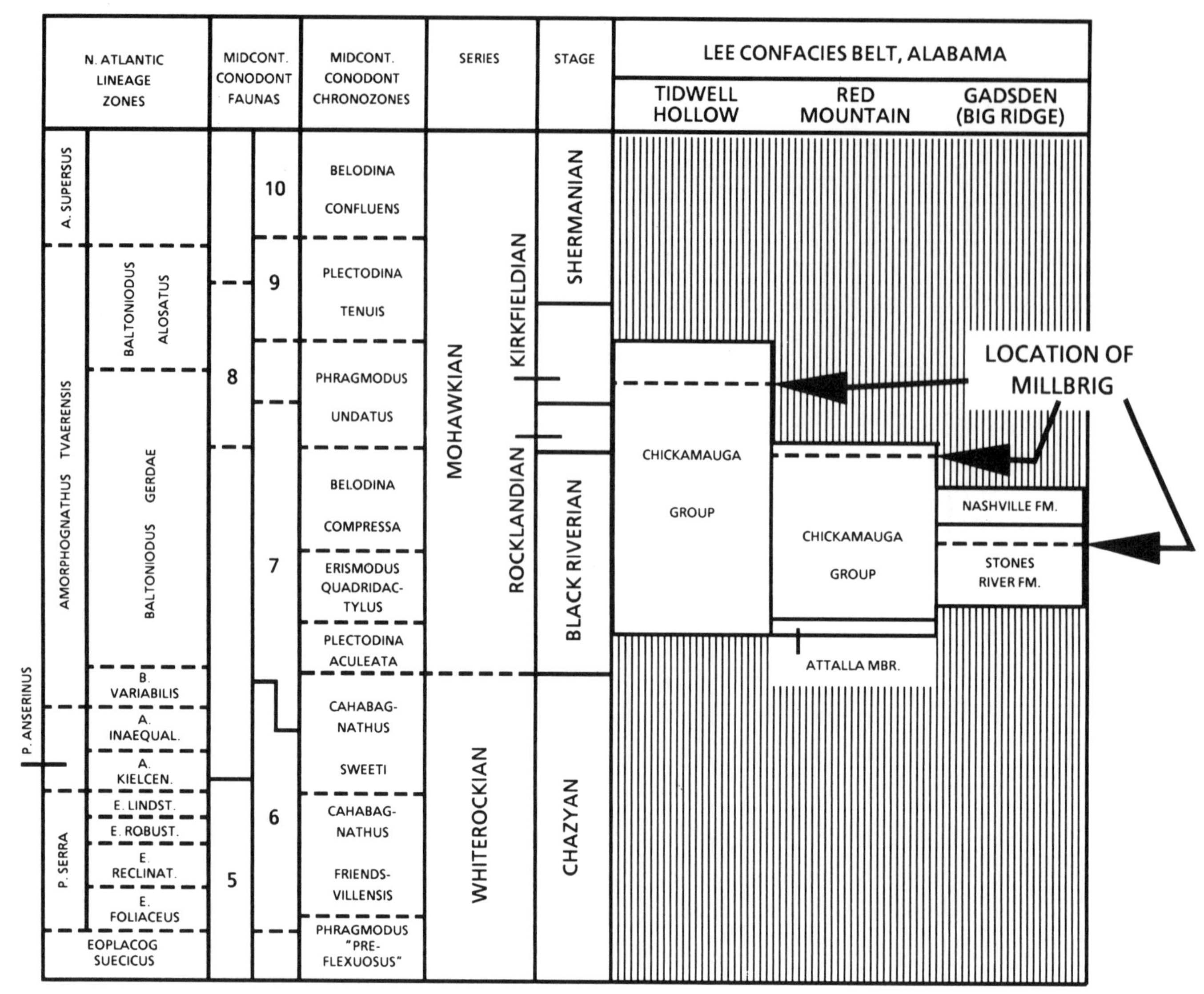

Figure 37. Conodont biostratigraphy of the Chickamauga Group limestones at the Tidwell Hollow, Red Mountain Expressway, and Big Ridge sections in Alabama, as adapted from Hall et al. (1986). The approximate location of the Millbrig K-bentonite at each section (arrows) is based on the results of the present study. The Millbrig, a true isochron, is obviously not time-transgressive as shown, even taking into consideration the approximation of its positioning given the scale of this illustration. Throughout much of the eastern midcontinent the Deicke and Millbrig are in the upper half of the *Phragmodus undatus* chronozone (Huff and Kolata, 1990), but as the platform facies (Tidwell Hollow) passes laterally into the shelf margin facies (Big Ridge) and then into the slope facies (Red Mountain), the Millbrig may in fact cross the lower boundary of that zone. The implication is that this conodont chronozone is diachronous across depositional strike in the southern Appalachians, a possibility that can be investigated by careful resampling of appropriate measured sections above and below the regionally correlatable K-bentonites.

Tidwell Hollow in the Murphrees Valley of Alabama (Figs. 2C and 31). There, the Carters-Hermitage sequence is readily identifiable even though the Carters and Hermitage are not formally recognized in that region. And the Millbrig occurs at the contact between those two lithofacies (Fig. 2C), just as it does in exposures along the Cincinnati Arch. As a result, the lithostratigraphy at Tidwell Hollow is nearly identical to that at the Tillet Brothers quarry in the Central Basin of Tennessee about 180 km to the north (Fig. 2B). The exposures near Tidwell Hollow in the Murphrees Valley are in the westernmost outcrop of Ordovician strata in the Alabama Valley and Ridge east of the Sequatchie Valley, and thus the Tidwell Hollow section, like the Hagan section farther north, is an important link in the correlation of the K-bentonite sequences of the Cincinnati Arch and the Valley and Ridge.

Correlation along strike in the Valley and Ridge

Previous studies. Rosenkrans (1936), Fox and Grant (1944), Huffman (1945), and Hergenroder (1966, 1973) showed that certain Ordovician K-bentonite beds could be correlated along strike for varying distances among sections in the southern Valley and Ridge. These and a very few other studies give detailed descriptions of the K-bentonites and their occurrence in various measured sections. Many attempts at correlation in this region have been put forth, and the following is a summary of some of these. Haynes (1992) presents a more detailed discussion of K-bentonite correlations in much of this stratigraphically and structurally complex region.

Rosenkrans (1936) found that beds V-3, V-4, and V-7 are the thickest K-bentonites in the central belt of Virginia, and that others in the section were much more difficult to recognize consistently from one exposure to the next because none were nearly as thick as those three. Rosenkrans was able to correlate several sections along strike in the central belt from the New River to Tazewell County, Virginia, a distance of about 100 km, and that correlation is supported by my findings. Bates (1939) studied several Ordovician sections in the north end of the Powell Valley in Virginia, and noted two thick K-bentonites in Turkey Cove, an area where the lithostratigraphy is very similar to the sequence at Hagan. Fox and Grant (1944) correlated their sections in the Chattanooga area with the sections in the Powell Valley described by Bates (1939), and with some sections in the Central Basin of Tennessee. Several of the K-bentonites around Chattanooga are described in exceptional detail by Fox and Grant. Huffman (1945) described the stratigraphy of the Powell Valley, and he correlated the section at Hagan with the Tazewell section of the central belt (Rosenkrans, 1936) and with exposures in the Bluegrass Basin of central Kentucky. Miller and Brosgé (1954) and Miller and Fuller (1954) described the stratigraphy of the Powell Valley in Virginia, and correlated K-bentonites among several sections in that region. Allen and Lester (1957) described the Ordovician sequence of northwest Georgia and presented correlations of several sections in the Georgia Valley and Ridge, many of which relied partly on the occurrence of a thick K-bentonite with an underlying chert layer. Only one such bed was noted by those authors, however, and because they used numbers rather than names to distinguish between their recognized geologic units (the K-bentonite was their bed "zero"), precise correlation with present-day units is difficult. Hergenroder (1966, 1973) studied the Bays Formation over its entire area of outcrop. Because of the many sections that he measured and described, the accuracy and detail of which are exceptional, his research was particularly important to the present study. Although Hergenroder's study focused on the stratigraphy and petrography of the Bays, he noted the stratigraphic importance of the various K-bentonites, and he suggested several along-strike correlations in the Bays as well as several across-strike correlations with the K-bentonites in the limestones in the Valley and Ridge farther west. The two thick K-bentonites that occur in some outcrops of the Bays in southeast Tennessee and northwest Georgia (Colvin Mountain Sandstone of Chowns and Carter, 1983a, and Carter and Chowns, 1986) were correlated with the two thick beds at Hagan, R-7 and R-10 of Miller and Fuller (1954) (Hergenroder, 1966, Plate 12). Hergenroder noted that the silicified zones which commonly occur at the base of the K-bentonites are more difficult to identify in the clastic sediments of the Bays, making correlations based on the presence of chert layers beneath the thicker K-bentonites impractical. Drahovzal and Neathery (1971), Chowns and McKinney (1980), Chowns and Carter (1983a), Lee (1983), Ward (1983), Jenkins (1984), and Carter and Chowns (1986) discussed the Rocklandian sequence of Alabama and northwest Georgia. Those studies all mention several K-bentonites in the Chickamauga Group limestones of the western belt, including T-3 and T-4, which were used as time lines in some of the correlations therein.

Discussion. In the southern Valley and Ridge province, the Deicke and Millbrig K-bentonites can now be correlated along strike from Alabama to Virginia. This is an area where the Ordovician outcrop can be divided into three distinct belts delineated by thrust fault traces and the facies changes that occur across them (Kreisa, 1980; Rader, 1982; Carter and Chowns, 1986; Haynes, 1992), and by lateral distribution of North Atlantic and Midcontinent conodont elements (Janusson and Bergström, 1980). These are the western, central, and eastern belts of Haynes (1992), the names used herein, and the Lee, Tazewell, and Blount confacies belts of Janusson and Bergström (1980).

The northern end of the western belt is the Powell Valley of Virginia and northeast Tennessee, where the Deicke and Millbrig occur in the Eggleston Formation. The stratigraphy of the Hagan, Harrogate, and Hinds Creek Quarry sections is very similar, with the Deicke occurring in the Eggleston 8 to 10 m below the Millbrig, which itself is 2 to 3 m below the Eggleston-Trenton contact (Fig. 27). The Deicke at Hagan (Fig. 5A), is the same K-bentonite shown in the center photograph of Figure 4 of Huffman (1945), who (incorrectly) identified the bed as K-bentonite V-4 of Rosenkrans (1936) (Haynes, 1992).

The Eggleston in the Powell Valley is lithologically very similar to the Tyrone and Carters Limestones of the Cincinnati Arch outcrop, and it is in the area of Clinton, Tennessee (Hinds Creek Quarry section) that the change in usage occurs from the Powell Valley nomenclature of Miller and Brosgé (1954) to the Cincinnati Arch nomenclature of Wilson (1949), which Milici (1969), Milici and Smith (1969), and Hatcher et al. (1992) found appropriate for use in the western Valley and Ridge of Tennessee and Georgia. Except for these differences in formation names, the stratigraphic sequence is otherwise little changed. Thus, in Tennessee west of the Kingston fault between Oak Ridge and Chattanooga, and in Lookout Valley and around Chickamauga in Georgia, the K-bentonites are still observed to occur at the top of a sequence of micrites (Carters Limestone) that is immediately overlain by a sequence of grainstones and packstones (Hermitage Formation). The Deicke is at the contact of the lower and upper Carters Limestone (Fig. 4A), with the Millbrig 6 to 8 m upsection, at or just beneath the contact of the Carters and the Hermitage (Fig. 4D). This stratigraphy is seen at the Henson Creek, Davis Crossroads, and Rising Fawn sections (Figs. 26 and 27).

In northeast Alabama from about the Georgia state line to Gadsden, the individual formations of the Stones River become increasingly less recognizable as one traces the K-bentonites toward the southwest. Thus the K-bentonites are assigned to the Stones River Group undivided, but in this region as in Tennessee and Georgia the Deicke occurs in Carters-like micrites (Fig. 3A) several meters below the Millbrig, which is found a few meters below the Stones River–Nashville contact, a widely recognized horizon equivalent to the Carters-Hermitage contact to the north and northeast. This stratigraphy is seen at the Ft. Payne, Big Ridge, and Bruton Gap sections (Fig. 27). At the Big Ridge section, the Deicke and Millbrig are only 3.3 m apart (Fig. 3B). This is the thinnest known interval of strata between the two beds in the study area. Only at certain exposures in the Upper Mississippi Valley do the Deicke and Millbrig occur this close or closer together (Kolata et al., 1986). This suggests that carbonate sedimentation rates in the Big Ridge area were lower than elsewhere in the Valley and Ridge and Cincinnati Arch. The localized uplift that was probably beginning to occur in the Birmingham area, an event that farther south had an appreciable effect on the thickness and lithofacies trends of Rocklandian and younger strata (Bearce, 1973), may have had a similar effect on sedimentation rates farther north as well.

The next sections to the south are in Birmingham, the Red Mountain Expressway and Green Springs sections (Fig. 27). These are very different lithologically from those at Ft. Payne, Big Ridge, and Bruton Gap, sections in which the lithostratigraphy of northwest Georgia (Carters to Hermitage) is still somewhat identifiable. In the Birmingham area there is no recognizable lithologic break between Stones River and Nashville lithofacies and, as a result, the strata between the Attalla Chert Conglomerate Member and the Sequatchie Formation are assigned to the Chickamauga Limestone undivided (Drahovzal and Neathery, 1971). At the Red Mountain Expressway section, much of the Chickamauga consists of Nashville-like fossiliferous nodular and planar bedded wackestones and packstones containing an abundant, open marine fauna, including small bioherms (Chowns and McKinney, 1980). These subtidal lithologies were deposited in a moderately deep outer shelf setting or even in an inner ramp setting seaward of the shelf margin (Benson, 1986), and they occur throughout the Chickamauga, above and below the Deicke and Millbrig. The thinness (80 to 85 m) of the entire Chickamauga at the Red Mountain Expressway section contrasts with the much thicker sections in northwest Georgia (over 440 m), where several tens of meters of Upper Ordovician strata (Sequatchie Formation and equivalents) occur between the K-bentonites and the Silurian Red Mountain Formation (Milici and Wedow, 1977; Carter and Chowns, 1986).

Significant differences occur between the Green Springs and Red Mountain Expressway sections themselves (Lee, 1983). Those exposures are only 4.5 km apart and yet the K-bentonite sequences are remarkably different. There is no lithostratigraphic evidence for an unconformity below the Millbrig, and so the absence of the Deicke at Green Springs (Fig. 27) suggests complete reworking by waves or currents (storm generated?) subsequent to deposition. The wave/current energy in this outer shelf or shelf margin environment is in marked contrast with the lower energy levels present in the peritidal environments that are inferred from the exposures in the Gadsden area and to the northeast, toward Georgia. Although the Deicke ash was evidently completely reworked at Green Springs, the Deicke is present as a discrete K-bentonite bed at the Red Mountain Expressway section not far away. It is, however, noticeably thinner than at sections farther northeast (Figs. 3A, B, and 27). At the Tidwell Hollow section north of Birmingham, the Deicke is again absent (Fig. 31), and the strata below the Millbrig (Fig. 2C), although containing some supratidal sediments, also contain higher energy sediments such as ooid grainstones and flat-pebble conglomerates. These observations suggest that significant reworking of the Deicke ash occurred at Tidwell Hollow, just as at Green Springs and the Red Mountain Expressway.

Whereas at Green Springs the Millbrig is a single thick bed of nearly normal thickness with an underlying layer of thick black chert, at the Red Mountain Expressway the Millbrig is really more a zone of bentonitic material and limestone almost 3 m thick, and the underlying chert layer is thin or absent. Biotite is present in moderate abundance throughout this zone (Fig. 3C, D), which is the interval from which the biotitic samples 1-11 to 1-16 of Raymond (1978), who first noted the abundant biotite in these sediments, were collected. Evidently much of the original ash was extensively reworked into the overlying calcareous sediment.

The presence of the Deicke and Millbrig in the fossiliferous Nashville-like limestones of the Chickamauga in the Birmingham sections implies a vertical distribution of this lithofacies that is in sharp contrast with all other sections in the western Valley and Ridge from the Powell Valley to Gadsden. The poorly fossiliferous subtidal to supratidal limestones of the Stones River Group, the Carters Limestone, and the Moccasin and Eggleston Formations, the units to the north and northeast of Gadsden with which the Deicke and Millbrig are associated and which represent deposition in an inner platform setting, clearly are *time-equivalent* to the more fossiliferous and normal marine limestones with which the K-bentonites are associated at Birmingham. It is the abundantly fossiliferous limestones of the Hermitage Formation and Trenton Limestone, however, which are the *facies-equivalents* of the Chickamauga Limestone at Birmingham. The recognition and identification of individual K-bentonite beds has made it possible to recognize these stratigraphic relationships, thus resolving one of the long-standing correlation problems in the Ordovician of the southern Appalachians.

In the central belt of the Valley and Ridge the lithologic character of Rocklandian strata differs markedly from that in the western belt (Haynes, 1992), but the Deicke and Millbrig are recognizable on the basis of their nonclay mineralogy. The along-strike correlations shown in Figures 28 and 29 are essentially the same as those first described by Rosenkrans (1936). Rosenkrans's detailed descriptions of the individual K-bentonite beds combined with the results of the present field and petrographic study are sufficient to identify bed V-3 of Rosenkrans as the Deicke and bed V-4 of Rosenkrans as the Millbrig (Haynes, 1992). In exposures on the St. Clair and western Narrows thrust sheets (Fig. 28), the Deicke occurs in the Eggleston Formation several meters above the uppermost redbeds of the Moccasin Formation, but in the more easterly exposures on the Narrows thrust sheet and on the Copper Creek thrust sheet farther south (Fig. 29) the Deicke occurs in the redbeds of the upper 2 to 3 m of the Moccasin Formation (Fig. 5B). The Millbrig occurs in the Eggleston Formation 6 to 8 m upsection from the Deicke throughout the central belt (Fig. 5C). The K-bentonite bed V-7 of Rosenkrans (1936), several meters upsection from the Millbrig, can also be found at many exposures (Haynes, 1992).

The Walker Mountain Sandstone Member (Butts and Edmundson, 1943; Haynes and Goggin, 1993) is present in the Moccasin Formation redbeds in the New River Valley. It is a 2- to 4-m-thick unit consisting of a basal conglomeratic quartz arenite overlain by greenish gray lithic sandstones and siltstones. In the central belt, the Deicke occurs in or immediately above the Walker Mountain at the Mountain Lake Turnoff, Goodwins Ferry, Gap Mountain, and Trigg sections, and the Millbrig occurs 6 to 8 m farther upsection. The stratigraphic relationships of the K-bentonites and the Walker Mountain Sandstone Member are discussed in more detail by Haynes (1992) and Haynes and Goggin (1993). Except for the presence of the Walker Mountain Sandstone Member in the New River Valley, the stratigraphy in the central belt changes little from the Thorn Hill section in Grainger County, Tennessee, northeast along strike to the Goodwins Ferry and nearby sections in Giles County, Virginia. North of Giles County there is a significant facies change beneath the Walker Mountain Sandstone Member, and the redbeds of the Moccasin pass laterally into the Eggleston and Nealmont Formations and other nonred units (Woodward, 1951; Kay, 1956; Lesure, 1957).

No central belt sections south of Eidson are included in Figure 29 because at the Lee Valley and Thorn Hill sections in Tennessee, the K-bentonite beds and the adjacent Moccasin Formation are very weathered and structurally disrupted. The stratigraphy at those sections is little different than at the Eidson section except that at Thorn Hill the uppermost Moccasin redbeds are above the Millbrig. The Eggleston in this area is extremely thin and it is not a mappable unit in the Clinch Mountain belt of Tennessee; at Thorn Hill there are only 2 to 3 m of Eggleston below the Trenton Formation.

South of Knoxville, the Moccasin-like lithologies that occur between the Clinchport and Kingston faults in Tennessee and northwest Georgia (Chowns, 1986) are very poorly exposed. In Alabama, the central belt stratigraphic sequence is very unlike that in the central belt of Virginia, possibly because of greater shortening along each major thrust fault in Alabama compared to the major thrust faults in southwestern Virginia. The Colvin Mountain Sandstone of the Alabama central belt is more similar compositionally to the Bays Formation, especially to the Walker Mountain Sandstone Member, than to the Moccasin Formation. The Colvin Mountain, like the Bays, is part of the Blount clastic wedge; therefore, discussion of the K-bentonites in the Colvin Mountain Sandstone is included in the following description of the eastern belt sequence.

The K-bentonites occur in the Blount Group, which is restricted to the central and eastern belts; these can now be correlated from Alexander Gap, Alabama, to Daleville, Virginia, a distance of about 700 km (Fig. 30). *This is nearly the entire lateral extent of the Blount clastic wedge.* Although Drahovzal and Neathery (1971) described two K-bentonites in the Colvin Mountain Sandstone at the Greensport Gap section in Alabama, I found only one K-bentonite at that exposure and, based on the presence of abundant euhedral ilmenite grains identified it as the Deicke. At Alexander Gap, about 20 km northeast of Greensport Gap along strike, both the Deicke and Millbrig are present (Fig. 18). The K-bentonite on Horseleg Mountain in Georgia was identified by Chowns and Carter (1983a) and Carter and Chowns (1986) as the T-4 K-bentonite of Wilson (1949), based on the alleged presence of biotite, but I found no biotite (or quartz) in that bed, only ilmenite or the pseudomorphic TiO_2 minerals after ilmenite, and authigenic K-feldspar. The presence of the Fe-Ti minerals indicates that this bed is the Deicke, not T-4, which is the Millbrig (Table 7, sample GA:FL 1-1).

The preservation of a homogenous bed of altered volcanic ash in a sequence of mature to supermature quartz arenites like the Colvin Mountain Sandstone is quite unusual, considering the wave and current energy of the depositional environment, a nearshore shallow marine setting (Jenkins, 1984). At Greensport Gap, the discrete bed of bentonitic material that contains abundant euhedral ilmenite grains and is therefore clearly recognizable as the Deicke, is only 8 to 10 cm thick (Fig. 4B). The matrix of that bed is a plastic blue-green bentonitic clay that is a mixed-layer illite/smectite with some kaolinite (sample AL:ET 1-2 in Table 4). The thinness of that bed, the gradational upper contact, and the presence of bentonitic clay in the overlying quartzose sands suggest that most of the ash was reworked into the overlying sands. Even the thin bed of bentonitic material contains rounded and clearly detrital terrigenous quartz sand grains, however, indicating that it has been reworked to some extent. Above the Deicke are cross-bedded, loosely consolidated quartz arenites with a bentonitic component. The cross-bedding, visible in the upper half of Figure 4B, is evidence of appreciable current activity that resulted in the reworking of most of the Deicke ash. Farther north, the absence of the Deicke in the eastern belt of Virginia and Tennessee is attributed to reworking in the conglomeratic sands of the Walker Mountain Sandstone Member (Haynes, 1992), so preservation of both beds in the Colvin Mountain Sandstone is quite unusual. Beall and Ojakangas (1967) described the occurrence of a K-bentonite in the Upper Cambrian Lamotte Sandstone of Missouri which, like the Colvin Mountain Sandstone, is largely a mature quartz arenite. Preservation of a discrete layer of volcanic ash in such a depositional setting seems to be very rare, however, and I am unaware of other reported occurrences of K-bentonites in a mature quartz arenite sequence.

The Greensport Formation and Colvin Mountain Sandstone extend discontinuously from Alabama eastward to the Armuchee Ridges of Georgia, where they are part of the Blount Group outcrop on the Clinchport thrust sheet, between the Clinchport and Rome Faults. In that region these strata were previously assigned to the Bays and Moccasin Formations (Butts and Gildersleeve, 1948; Allen and Lester, 1957; Hergenroder, 1966; Cressler, 1970; Chowns and McKinney, 1980), the type areas of which are in northeast Tennessee and southwest Virginia. Because of the distance to the type areas, Chowns and Carter (1983a) suggested that the nomenclature from nearby sections in Alabama is more appropriate for Georgia, and that usage is followed herein. Although other geologists have reported K-bentonites from several exposures in those outcrop belts (Allen and Lester, 1957; Hergenroder, 1966; Cressler, 1970), the only K-bentonite found during the present study was on Horseleg Mountain near Rome (Fig. 30). That bed, which occurs in the uppermost Greensport Formation immediately below an approximately 1-m-thick quartz arenite bed that is the basal unit of the Colvin Mountain Sandstone (Chowns and Carter, 1983b), contains abundant euhedral ilmenite grains, and on this basis it is identified as the Deicke. Kaolinized, authigenic K-feldspar grains are also present.

North of Horseleg Mountain, the outcrop of the Greensport and Colvin Mountain Formations in the Armuchee Ridges continues and, although Rosenkrans (1936), Allen and Lester (1957), and Hergenroder (1966) reported K-bentonites from near Rocky Face, Georgia, no exposures containing K-bentonites were found in that region.

In eastern Tennessee, K-bentonites occur in the Bays Formation in the foothills region of the Great Smoky Mountains, and Rodgers (1953) and Hergenroder (1966) described several exposures of the Bays that contain K-bentonites. The Citico Beach section southwest of Chilhowee Mountain and the Little Tennessee River (Fig. 30) is the best exposure that I have found in this area. The stratigraphy of this section is in some ways very much like that at Horseleg Mountain, in Georgia, but it is very different from the sections along strike to the north in the Bays Mountains of northeast Tennessee. At the Citico Beach section only the Deicke is present. It is the uppermost bed of the Bays, it is almost 5.5 m thick, and it is immediately overlain by a dark brown shale that has historically been assigned to the Chattanooga Shale of Devonian-Mississippian age (Rodgers, 1953; Neuman, 1955). Because no black shale is present, and because the Chattanooga-Grainger contact is transitional, the unit immediately overlying the Deicke is most likely either the Brallier Member or the Big Stone Gap Member rather than the Millboro Member of the Chattanooga Shale (Hasson, 1985).

This sub-Chattanooga unconformity is traceable throughout the eastern Valley and Ridge from Virginia to Alabama. In southwest Virginia and northeast Tennessee, the Chattanooga Shale unconformably overlies the Wildcat Valley Sandstone of Early and Middle Devonian age (Hasson, 1985; Dennison et al., 1992). In the eastern Valley and Ridge of Alabama, the Chattanooga Shale unconformably overlies the Devonian Frog Mountain Sandstone, a unit similar to the Wildcat Valley Sandstone, and in places it overlies the Silurian Red Mountain Formation. Elsewhere in eastern Alabama the Chattanooga is absent and the Lower Mississippian Maury Shale or Fort Payne Chert unconformably overlies the Lower Ordovician Knox Group (Thomas, 1982). It should be noted that the lowest 2 to 3 m of the Chattanooga Shale are faulted and otherwise very structurally contorted at the Citico Beach section.

In Virginia and northeast Tennessee, K-bentonites occur in the Bays Formation east of the Saltville fault, and east of the Pulaski fault in and north of the New River Valley. It is in the region between the Citico Beach section and the Terrill Creek section that the Deicke disappears from the Bays (Fig. 35) and from Terrill Creek northward the Millbrig and K-bentonite V-7 of Rosenkrans (1936) are the most prominent K-bentonites in the Bays (Haynes, 1992). And unlike the extremely altered K-bentonites collected from the Colvin Mountain Sandstone in Alabama and Georgia, the Millbrig can be recognized in the Bays by a careful examination of hand samples, which reveals the presence of abundant biotite. Thin sections of samples from several exposures show that this biotite is appreciably kaolinized and chloritized (Fig. 10D).

As shown in Figure 30, the thick K-bentonite in the Bays Formation at the Daleville, Connor Valley, Crockett Cove, Chatham Hill, and Terrill Creek sections is identified as the Millbrig (Figs. 4C and 5D). At those five sections the 18- to 28-m-thick interval between the Walker Mountain Sandstone Member and the Millbrig is completely exposed. In addition, much of the Bays below the Walker Mountain is also exposed at the Chatham Hill and Nebo sections, and no K-bentonites were observed in that interval. At the Crockett Cove, Connor Valley, and Daleville sections, the Walker Mountain Sandstone Member is the basal unit of the Bays. It unconformably overlies the Witten Formation (Crockett Cove), Wassum Formation (Connor Valley), and the Liberty Hall Formation (Daleville) (Haynes, 1992). The first thick K-bentonite above the Walker Mountain Sandstone Member in those sections is identified as the Millbrig on the basis of its nonclay minerals. Most notably it contains abundant biotite that is chloritized to varying degrees but is still recognizable as biotite.

In most sections of the Bays where the Millbrig occurs, it is a thick, biotite-rich K-bentonite, and although it is grayish red at Chatham Hill, it is grayish white with red mottles at the Terrill Creek section, and yellow to greenish gray at the Crockett Cove, Connor Valley, Catawba, and Daleville sections. Those differences reflect the degree to which the surrounding sediments were oxidized during diagenesis. Biotite in Millbrig samples from the Crockett Cove, Catawba, and Daleville exposures is bronze to black in color, rather than the blue-green color of the chloritized biotites at Chatham Hill and Terrill Creek. As a result, the samples from those three sections are texturally very similar to Millbrig samples collected from the central belt. The Millbrig at Connor Valley is very weathered, but bleached biotite is visible in hand samples. Quartz and feldspar phenocrysts are generally rare in Millbrig samples from the Bays, but they are relatively abundant at Daleville, although the feldspars are highly altered. Also, the sandstones overlying the bed at Crockett Cove contain abundant biotite both along bedding planes and throughout the rock in the 2 to 3 m immediately above the K-bentonite (Fig. 30).

Except at the Citico Beach section, the Deicke is absent from the Bays Formation, which is restricted in occurrence to the eastern belt. The Deicke is present in the central belt just a few kilometers to the northwest, and at the Trigg, Goodwins Ferry, and Mountain Lake Turnoff sections, it occurs immediately above the conglomeratic bed of the Walker Mountain Sandstone Member. In the eastern belt, therefore, the Deicke ash was evidently completely reworked and mixed in with those conglomeratic sands, which thicken toward the east (Haynes, 1992). The Deicke does not contain biotite or another dark mineral readily apparent in hand samples, so it is difficult to determine visually whether or not any reworked ash is present in the surrounding strata, as is possible with the Millbrig based on the presence of what is clearly reworked biotite at the Crockett Cove, Catawba, and Red Mountain Expressway sections. This problem could be investigated by petrographic analysis to determine whether an unusual concentration of ilmenite or its derivatives is present in the part of the Walker Mountain Sandstone Member, or to identify some unique geochemical signature of ilmenite in known Deicke samples that might match the same geochemical signature of ilmenite in the Walker Mountain.

Correlation across strike in the Valley and Ridge

Previous studies. Very few authors have suggested correlations of Rocklandian K-bentonites across strike in the Valley and Ridge province, probably because of the stratigraphic and structural complexity of that region. In Alabama and Georgia, the T-3 and T-4 K-bentonites in the Chickamauga Group limestones of the western belt have been correlated with two K-bentonites in the Colvin Mountain Sandstone at Greensport Gap, Alabama (Drahovzal and Neathery, 1971; Chowns and McKinney, 1980; Jenkins, 1984), and T-4 has been correlated with a K-bentonite in the Colvin Mountain Sandstone at the Horseleg Mountain and Dalton sections in Georgia (Chowns and Carter, 1983a; Carter and Chowns, 1986). The above authors' justification for identifying T-4 in the Colvin Mountain sections was the reported presence of small grains of biotite in the upper K-bentonite at Greensport Gap, and abundant biotite in the bed at Horseleg Mountain and Dalton. The lower bed, where present, was then correlated with T-3. They also supported those correlations by noting the presence of a 0.05-m-thick quartz arenite bed between the T-3 and T-4 K-bentonites at the Big Ridge section in Alabama (Chowns and McKinney, 1980), a stratigraphic positioning which suggests that it is a very thin tongue of the Colvin Mountain Sandstone, the unit in which K-bentonites occur across the Helena fault.

In Virginia, Huffman (1945), Miller and Fuller (1954), and Hergenroder (1966, 1973) all believed that K-bentonites V-4 and V-7 of the Tazewell section of Rosenkrans (1936) (the Plum Creek section of the present study) correlated with the two thick K-bentonites in the Powell Valley at Hagan named R-7 and R-10 by Miller and Fuller (1954). In addition, Rosenkrans (1936) and Hergenroder (1966) believed that V-3 correlated eastward from the central belt with a thick K-bentonite in the Bays Formation near Roanoke, in the eastern belt.

Discussion. Correlation of the K-bentonites across strike in Alabama is presented in Figure 31. This cross section includes the position of the Helena thrust fault relative to the sections shown. The Millbrig is recognizable on the basis of its phenocryst assemblage at Tidwell Hollow and Big Ridge, and it is probable that the upper of the two thick K-bentonites at Alexander Gap is the Millbrig. The presence of ilmenite and the absence of biotite or quartz in the lower K-bentonite at Alexander Gap indicate that that bed is the Deicke, hence the upper bed is most logically the Millbrig. The thin quartz arenite between the Deicke and Millbrig at the Big Ridge section is shown (Fig. 31), and as noted by Chowns and McKinney (1980), this thin Colvin Mountain–like sandstone reinforces the correlation of the two K-bentonites in the Colvin Mountain Sandstone at Alexander Gap with the Deicke and Millbrig. In Georgia, the equivalent across-strike correlation would be between the Rising Fawn section (Fig. 27) and Horseleg Mountain section (Fig. 30).

In Virginia and Tennessee, the K-bentonite correlations suggested by Huffman (1945), Miller and Fuller (1954), and Hergenroder (1966, 1973) have recently been discussed in detail and in part reinterpreted (Haynes, 1992). It is not V-4 and V-7 of Rosenkrans (1936) that correlate with the two thickest K-bentonites at Hagan, R-7 and R-10 of Miller and Fuller

(1954), but instead it is V-3 and V-4 that correlate with those two thick K-bentonites at Hagan, which are the Deicke (R-7) and Millbrig (R-10). V-7 is younger than those, and is equivalent to bed R-12 of Miller and Fuller (1954). Bed V-3 of Rosenkrans (1936) is the Deicke, and in Virginia it does not persist eastward into the Bays Formation, but the Millbrig and bed V-7 do occur in the Bays.

Figures 32 and 33 show correlations between the western, central, and eastern belts in Virginia and Tennessee. These correlations are between strata on separate thrust sheets, and the positions of the various major thrust faults crossed by these lines of correlation are shown between the appropriate sections. As the K-bentonites are traced across the various thrust sheets to the east, the Eggleston-Trenton contact becomes the Bays-Trenton contact, as seen at the Terrill Creek section (Fig. 32). That contact can be readily identified throughout the study area because it marks the sudden appearance of the fossiliferous grainstones and packstones that characterize the Trenton Formation. It is most important in southwesternmost Virginia and northeast Tennessee because in that area two of the beds that are important markers farther north are absent: the Walker Mountain Sandstone Member is not present in the central belt, and bed V-7 was not observed in the Bays (Fig. 32). Nevertheless, in the vicinity of the Terrill Creek section the Bays-Trenton contact is recognizable (Hergenroder, 1966), so even though the Walker Mountain Sandstone Member and bed V-7 are of no use for correlation purposes, the correlation shown in Figure 32 between Eidson and Terrill Creek is supported by the lateral persistence of that contact.

Farther north in Virginia, the Millbrig and bed V-7 can be correlated across both the Saltville and Pulaski faults as shown in Figure 33. The Millbrig is nearly parallel to the Walker Mountain Sandstone Member (Fig. 30), a unit that is several meters downsection and is an important marker bed in this region (Haynes, 1992; Haynes and Goggin, 1993). At Daleville the Deicke is absent and the Millbrig and V-7 K-bentonites occur in the green Bays Formation. In the Moccasin Formation at the Goodwins Ferry, Trigg, and Mountain Lake Turnoff sections on the Narrows thrust sheet, the Deicke is present, and the occurrence of 2 to 3 m of green Bays-like sandstones in the upper Walker Mountain Sandstone Member several meters below the Millbrig and immediately above the Deicke reinforces this correlation (Haynes, 1992). The K-bentonite stratigraphy is thus recognizable at each of these sections, and the line of correlation crosses both the Saltville and the Pulaski faults.

STRATIGRAPHIC VALUE OF THE DEICKE AND MILLBRIG

Introduction

Now that the Deicke and Millbrig can be recognized and correlated throughout the southeastern U.S., their potential as time planes can be better realized. For example, on a local scale, at the Red Mountain Expressway in Birmingham the Millbrig is only 3 to 4 m below the angular unconformity that separates the Chickamauga Limestone from the overlying Sequatchie and/or Red Mountain Formation, whereas to the north and northeast the Millbrig is several tens of meters below the Sequatchie. Recognition of the Deicke and Millbrig in the uppermost Chickamauga confirms that almost the entire Nashville Group as defined elsewhere in Alabama is absent in Birmingham. Therefore, instead of referring to the strata between the Attalla Chert Conglomerate Member and the Red Mountain Formation simply as the Chickamauga Limestone, as is now done (Chowns and McKinney, 1980), I suggest that those strata could be assigned to the Stones River Group. The Stones River is recognized in northeastern Alabama (Drahovzal and Neathery, 1971), and at Birmingham the presence of Stones River–age strata is confirmed by the K-bentonite stratigraphy. The significant lithologic differences between the Birmingham sections and those northeast of Gadsden can now be the focus of some very significant sedimentologic studies on facies and faunal relations of the Ordovician sequence, instead of being a stumbling block that hinders a comparative study.

Because the Deicke and Millbrig are isochrons, an opportunity is also presented for the sedimentologic, paleontologic, and geochemical study of a very narrow but time-equivalent sequence of strata over a very wide area. Such a study of the stratigraphic interval sandwiched by both beds would be of great value given the rarity of being presented with an interval that is bracketed by two exceptionally well defined isochrons.

Two examples of just how important and useful the Deicke and Millbrig can be in local and regional litho- and biostratigraphic studies are presented in the following discussion.

Sequence stratigraphy and unconformities

The concept of sequence stratigraphy (Mitchum et al., 1977; Van Wagoner et al., 1988) has revolutionized the way geologists look at and think about sequences of strata. Sequence stratigraphy has provided a stimulus for field-oriented sedimentary geologists to undertake a critical reexamination of the stratigraphic succession in many intensely studied "classic" areas, like the southern Appalachians discussed herein. Before the fundamental units of sequence stratigraphy (sequences, parasequences, systems tracts, condensed sections, marine flooding surfaces) can be recognized, however, it is necessary to identify and describe carefully all unconformities and their correlative conformities in the stratigraphic interval(s) of interest. From this information both lithostratigraphic and chronostratigraphic charts can then be prepared for these intervals, and based on the identified unconformities the fundamental units can be recognized.

A detailed examination of the K-bentonites and the associated sequences in the Ordovician of this region is underway, and much additional information will be reported at a future time, but the following discussion summarizes the preliminary findings of this study that pertain to one of the most important steps in sequence stratigraphy, the identification of unconformities.

Although the structural disruption of the Valley and Ridge complicates recognition of subtle stratigraphic changes, boundaries, etc., a more serious problem is that the Middle and Upper Ordovician strata of this region record the transition from a trailing to a leading edge tectonic regime, an event that resulted in development of a foreland basin on downwarped continental crust at the craton edge, which in turn led to significant changes in regional sedimentation patterns (Read, 1980). Thus, because the original sequence stratigraphy concepts were developed for trailing edge settings marginal to an oceanic basin where eustatic sea-level changes, not tectonic activity, control sedimentation, the application of sequence stratigraphic principles to these Ordovician strata is not simply an exercise in identifying sequence bounding unconformities and then applying the terminology of Van Wagoner et al. (1988). Nonetheless, with study and recognition of changes in provenance and of localized unconformities, it should be possible to identify sedimentation events that were most likely tectonically driven, and events that were mostly the result of a eustatic sea-level change. Such work has been done for the Silurian and Devonian of southwest Virginia (Dennison et al., 1992), and the beginnings of a sequence stratigraphic framework for the Middle and Upper Ordovician can now be identified as well.

Throughout the Cincinnati Arch and much of the western Valley and Ridge, there are several recognized unconformities in the Paleozoic sequence, and these are labeled A through F in Figure 36, which includes two regional cross sections from the Cincinnati Arch to the eastern Valley and Ridge. Some are especially extensive and widely recognized: unconformity A is the post-Knox unconformity that is overlain variously by the Attalla Chert Conglomerate Member in Alabama, the Pond Spring Formation in Georgia, and the Mosheim, Blackford, and New Market Formations in Tennessee and southwest Virginia; unconformity D is the sub-Silurian unconformity that is overlain by the Red Mountain Formation in Alabama and Georgia, and the "upper" Tuscarora/Clinch Sandstone in Tennessee and Virginia; unconformity E is the sub-Devonian unconformity that is overlain variously by the Chattanooga Shale, Armuchee Chert, or Frog Mountain Formation in Alabama and Georgia, and the Chattanooga Shale, Huntersville Chert, or Oriskany Sandstone in Tennessee and extreme southwest Virginia, and unconformity F is the sub-Mississippian unconformity that is overlain by the Fort Payne Chert in Alabama and Georgia (Thomas, 1982; Hasson, 1985; Mussman and Read, 1986; Dennison et al., 1992). Although unconformity E can actually be a merger of several intra-Devonian unconformities at any one exposure (Dennison et al., 1992), for simplicity it is considered a single, sub-Devonian, unconformity herein because it is not a major part of the present discussion.

The existence of recognizable and correlateable Rocklandian K-bentonites provides a means of identifying and delineating the lateral and vertical extent of some lesser unconformities as well. In the Cincinnati Arch outcrop of Kentucky and Tennessee, the Millbrig occurs immediately beneath unconformity B_2, which is referred to herein as the post-Tyrone unconformity because of its occurrence above the Tyrone Limestone in central Kentucky (Fig. 24). This unconformity can be traced southward across the Cumberland Saddle into the central Tennessee outcrop, and then into the northwesternmost Valley and Ridge of Alabama (section A-A′ in Fig. 36). In all those areas the Millbrig occurs immediately beneath the break in section at the Carters Limestone–Hermitage Formation contact. In much of the western Valley and Ridge of Tennessee, Georgia, and Alabama, however, unconformity B_2 passes laterally into the conformable contact between the Stones River and Nashville Groups, the Tidwell Hollow section in Alabama being a notable exception (Fig. 3C). At Birmingham, where there are no typical Stones River lithofacies, the Stones River–to–Nashville transition is not recognized at all, a stratigraphy which indicates that unconformity B_2 is truly a conformity in much of the Alabama Valley and Ridge. In the exposures along Beaver Creek Mountain and Colvin Mountain east and northeast of Birmingham, the Deicke and Millbrig are present, and there is again no major stratigraphic break at the horizon of unconformity B_2. Instead, the K-bentonites occur in the quartz arenites of the Colvin Mountain Sandstone.

A locally significant unconformity is especially well developed at the Birmingham area sections. This is unconformity C, the post-Chickamauga unconformity, which is overlain at the Red Mountain Expressway by the thin Sequatchie Formation, or, where the Sequatchie is absent, by the Silurian Red Mountain Formation. As shown in cross section A-A′ of Figure 36 unconformity C represents a merging of several disconformities that are present to the north between the various units of the Nashville Group. To the northeast of Birmingham the magnitude of unconformity C rapidly diminishes, and in northwest Georgia there is no unconformity at this horizon (Chowns and McKinney, 1980). East and southeast of Birmingham, there is a merger of unconformities C, D, and E, and at a quarry near Ragland, the Ordovician Little Oak Limestone, with its Millbrig-like K-bentonite in the uppermost 4 m, is unconformably overlain not by Silurian strata, but by sandstones of the Devonian Frog Mountain Formation (Drahovzal and Neathery, 1971). In most sections east of Birmingham, however, *all* Middle and Upper Ordovician strata, and any K-bentonites therein, are absent because of the merging of unconformities A, C, D, E, and commonly F, and as a result Mississippian strata (Maury Shale and Ft. Payne Chert) overlie Lower Ordovician strata (Newala Limestone) (Thomas, 1982). Stratigraphic interpretations based on K-bentonite correlations cannot be made, and thus cross section A-A′ is not extended eastward beyond the position of the Eden (Coosa) fault.

To the northeast unconformities A, D, and E persist, and are present in southwest Virginia (cross section B-B′ in Fig. 36). In Kentucky unconformity B_2 is recognizable as the post-Tyrone unconformity immediately above the Millbrig; to the east it becomes the conformable contact between the Eggleston and Trenton Formations in the Valley and Ridge of southwest Virginia and Tennessee. Still farther east it is the conformable contact between the Bays and Trenton Formations (Kreisa, 1980). These stratigraphic relations mirror those in Tennessee, Alabama, and Georgia, thus confirming that unconformity B_2 is confined to the Cincinnati Arch and is not present in most of the Valley and Ridge.

Unconformity C of Alabama, the post-Chickamauga unconformity at Birmingham, is not present in Virginia, but another locally significant unconformity is present, and it too is closely associated with the Rocklandian K-bentonites. This is unconformity B_1, the sub–Walker Mountain Sandstone Member unconformity. This unconformity occurs beneath the conglomeratic quartz arenites of the Walker Mountain Sandstone Member (Hergenroder, 1966; Haynes, 1992; Haynes and Goggin, 1993), which is a member of both the Bays and Moccasin Formations (cross section B-B′, Fig. 36) and is the basal member of the Bays in the exposures where unconformity B_1 is most pronounced (the Crockett Cove, Connor Valley, Ellett, Daleville, and Peters Creek sections). In the Nebo and Chatham Hill sections, at the southwestern end of Big Walker Mountain, the Walker Mountain is conglomeratic but is within the redbeds well above the base of the Bays Formation (Butts and Edmundson, 1943; Haynes, 1992). In the New River Valley along Gap Mountain in the same strike belt farther north, where the Bays has passed laterally into the Moccasin Formation, the Walker Mountain and unconformity B_1 are present, having persisted through the facies change. To the east of Wytheville and Roanoke, no strata as young as Rocklandian are present, and stratigraphic interpretations based on K-bentonite correlations cannot be made. Thus the cross section is not extended eastward beyond the position of the Salem fault.

Although the major unconformities of the region are well known, the recognition of correlateable Rocklandian K-bentonites has facilitated understanding the stratigraphic relationships of unconformities B_1, B_2, and C, three less widespread unconformities now identified in the southern Appalachians. With recognition of these unconformities work can begin on delineating the sequence stratigraphic boundaries for the Ordovician depositional sequences in this region. The local tectonic history of the eastern margin of the Valley and Ridge will be better understood now that the nature and extent of unconformities B_1, B_2, and C are known.

This short discussion on the usefulness of the Rocklandian K-bentonites in determining the extent of several locally and regionally important unconformities demonstrates the potential usefulness of the Deicke and Millbrig in sequence stratigraphic studies of the Ordovician throughout the eastern midcontinent.

Comparison of K-bentonite lithostratigraphy with conodont biostratigraphy

Correlations based on the vertical distribution of certain fossils such as conodonts are extremely important in studies of Paleozoic strata. Although some fossils—especially conodonts and graptolites—are used by many geologists as de facto isochrons in Ordovician strata, their paleogeographic distribution cannot be expected to be indiscriminant of every biofacies change, unlike a single layer of volcanic ash, altered or unaltered. Therefore, the most useful correlation charts would be those on which vertical ranges of conodonts or graptolites are tied in to the position of recognizable marker beds of regional or even global extent. The present study and that of Haynes (1992) are the first to present a unified, region-wide stratigraphic framework for the most widespread Rocklandian K-bentonites of the southern Appalachians, with "decoder" charts for all of the older alpha-numeric nomenclature systems that were applied to the K-bentonite sequence. Thus it is worthwhile to compare the stratigraphic interpretations of the present study with an interpretation based on conodont biostratigraphy.

Figure 37, adapted from Hall et al. (1986), is based on a detailed study by Hall (1986) of conodont assemblages in the Chickamauga Limestone and its equivalents at three measured sections in Alabama. Based on a comparison of conodonts recovered from samples with the regional distribution of Ordovician conodont zones and taxa (shown in Fig. 37), Hall (1986) determined that the top of the Chickamauga is youngest at Tidwell Hollow, older at Red Mountain, and oldest at Gadsden (the Tidwell Hollow, Red Mountain Expressway, and Big Ridge sections, respectively, of the present study). To these three sections of Hall et al. (1986) I have added the approximate position of the Millbrig K-bentonite based on a comparison of my section measurements with those of Hall (1986). At Tidwell Hollow and Big Ridge the Millbrig occurs 15 to 25 m below the contact of the Chickamauga Group and the overlying Sequatchie Formation, but at the Red Mountain Expressway the Millbrig is only 3 m below that contact.

There are significant discrepancies between the findings of Hall et al. (1986), summarized in Figure 37, and those of the present report. First, it is not possible for the Millbrig to be Kirkfieldian in age at Tidwell Hollow, Rocklandian in age at the Red Mountain Expressway, and Blackriveran in age at Big Ridge, as is indicated by correlation based on conodonts. Throughout much of the eastern midcontinent the Deicke and Millbrig are in the *Phragmodus undatus* chronozone, as they are at Tidwell Hollow and the Red Mountain Expressway. Their occurrence in the *Belodina compressa* chronozone at Big Ridge is thus a clear indication that these zones are diachronous. Second, the conodont assemblages indicate that the Stones River and Nashville Groups at Big Ridge are entirely Blackriveran in age, but there is accumulating evidence that in much of the eastern midcontinental United States the Blackriveran-Rocklandian boundary—the "golden spike" arbitrarily designating the base of the Rocklandian—can and should be placed at the base of the Deicke K-bentonite (Kolata et al., 1986; Sloan, 1988; Huff and Kolata, 1990; Haynes, 1992). Because the Deicke is absent at Tidwell Hollow (the reason for using the Millbrig in Fig. 37), the base of the Rocklandian cannot be as accurately placed there using the K-bentonites, but it can be defined at both the Big Ridge and Red Mountain Expressway sections, where the Deicke is present. Third, the conodont assemblage of the upper 6 m at the Red Mountain section indicates that those strata are younger than the entire Nashville Formation at the Big Ridge section, although Hall et al., (1986) were unable to recognize identifiable conodonts in the uppermost Nashville at Big Ridge. Because no unconformity has been noted at the Red Mountain Expressway section in the Chickamauga below the Sequatchie (Hall et al., 1986), the only way to reconcile their interpretations with mine is to have a condensed section above the K-bentonites at Red Mountain. As noted above, the entire Chickamauga Limestone at Birmingham is much thinner than elsewhere in the region (Chowns and McKinney, 1980), and part of it may in fact be a condensed section *sensu* Loutit et al. (1988), but there is no evidence that only the 3 to 4 m of section above the K-bentonites is a condensed section. Therefore, the conclusion of Hall et al. (1986) that the entire 3 to 4 m of Chickamauga above the K-bentonites at the Red Mountain Expressway is younger than all of the Nashville at Big Ridge cannot be reconciled with the lithostratigraphy of the Red Mountain Expressway section. The Deicke and Millbrig K-bentonites are only 3 to 4 m below the angular unconformity that separates the Chickamauga Limestone from the overlying Sequatchie and/or Red Mountain Formation; furthermore, there is no lithostratigraphic evidence for any unconformities in the 2- or 3-m interval between the Millbrig and the unit overlying the angular unconformity.

Certainly one cause for the discrepancies between the findings of Hall et al. (1986) that are based on conodont biostratigraphy, and results herein based on K-bentonite lithostratigraphy, is that the post-Nashville–age conodonts at the Red Mountain Expressway all came from late Mohawkian or Cincinnatian strata assignable to the Sequatchie Formation. There are in fact significant differences of opinion in the literature concerning the age and stratigraphic assignment of the approximately 10 m of strata at the Red Mountain Expressway that occur above the angular unconformity 3 to 4 m upsection from the Millbrig (shown in Fig. 15 of Chowns and McKinney, and Fig. 27 herein) and below the oldest hematitic ironstone bed. Drahovzal and Neathery (1971), Chowns and McKinney (1980), and Thomas et al. (1982) place these 10 m or so of limestones and siltstones in the Silurian Red Mountain Formation; this is the stratigraphy used in Figure 27 herein. Brockman (1978) strongly disagreed with the assignment of these 10 m of strata to the Silurian, however. He instead favored assignment to either the Colvin Mountain Sandstone or the Sequatchie Formation. Although Brockman made this recommendation primarily from his consideration of lithologies, he also noted that the fauna, although ". . . sparse, poorly preserved, and somewhat ambiguous, . . . generally indicates Maysvillian and/or Richmond age" (Brockman, 1978, p. 6). Furthermore, conodonts in these strata suggest an Ordovician age, and as a result those geologists, including Hall, who have studied the conodont faunas have chosen to assign these strata to the Upper Ordovician (Raymond, 1978; Hall, 1986; Hall et al., 1986). This interesting stratigraphic problem, which is beyond the scope of the present work, deserves further study.

Another possible cause for the discrepancies is that because the inferred depositional environments of the Chickamauga in the Birmingham area are markedly different from those inferred at Tidwell Hollow and Big Ridge, the conodont assemblages might also be different. The relatively abundant and diverse fauna in the nodular packstones and wackestones of the Chickamauga at the Birmingham sections indicates deposition in a normal marine environment on the outer shelf or shelf margin, whereas at Tidwell Hollow and especially at Big Ridge, the sparse fauna in the mudcracked and predominantly micritic limestones suggests deposition in more restricted peritidal environments of the inner shelf (Chowns and McKinney, 1980). Mixing of conodont forms would not be unexpected in the deeper water nodular limestones of the Birmingham region, particularly if the conodonts from the Midcontinent and North Atlantic provinces really are just warm-water and cold-water faunas, respectively (Sweet and Bergström, 1974). In fact, the recovered assemblage at the Red Mountain Expressway section is a mixture of conodont forms from the two provinces (Raymond, 1978; Hall, 1986; Hall et al., 1986).

Janusson and Bergström (1980) quantified the conodont provincialism problem by dividing the southern Valley and Ridge into three belts characterized by distinct conodont faunas. These are the Lee Confacies Belt (western Valley and Ridge), which is characterized by Midcontinent faunas, the Blount Confacies Belt (eastern Valley and Ridge), which is characterized by North Atlantic faunas, and the Tazewell Confacies Belt (central Valley and Ridge), which contains a mixture of the Lee (90%) and Blount (10%) Confacies Belt conodonts. The Red Mountain Expressway section is thus in the Tazewell Confacies Belt (Hall, 1986).

Although the work by Janusson and Bergström (1980) greatly increased our understanding of the regional conodont distribution patterns, there are still many unresolved problems with the use of conodont biozones as time planes in correlations of Ordovician strata in the southern Valley and Ridge because the Midcontinent and North Atlantic conodont faunas are so different. In addition to the problems of matching North Atlantic and Midcontinent forms, there are important sedimentologic differences that have to be considered as well, and which may ultimately have a bearing on the conodont problem described above. As noted by Chowns and McKenny (1980), the Chickamauga Limestone at the Red Mountain Expressway is clearly sedimentologically more similar to the Nashville Group (nodular and wavy bedded limestones with a diverse and abundant open marine fauna interpreted as being deposited in a shelf to ramp setting) than to the Stones River Group (peritidal, restricted fenestral micrites with a sparse fauna), even though the K-bentonite stratigraphy shows the

Chickamauga Limestone to be the same age as the Stones River Group (Fig. 27).

My interpretation of this stratigraphy is that the fossiliferous nodular limestone lithofacies and the underlying peritidal lime mudstone lithofacies were transgressive, and thus diachronous, from Birmingham and areas farther south and east, toward the north and northeast, crossing the K-bentonites in that direction. The nodular limestone lithofacies surrounds both the Deicke and Millbrig at Birmingham, but at the Bruton Gap and Big Ridge sections (Fig. 27) to the northeast, the base of this lithofacies is the base of the Nashville Group, which occurs several meters above the Millbrig. Farther north this is the lithofacies of the Hermitage Formation, the lower Lexington Limestone, and the Trenton Formation, all of which are stratigraphically above the Deicke and Millbrig. *Possibly the conodont faunas were also diachronous to some extent in that direction as well.* Assuming the correct identification of K-bentonites in the various measured sections presented herein, the lithostratigraphy makes this interpretation compelling because conodont biozones and subzones, although cutting across many lithofacies and other biofacies, are not indiscriminant of every change in environmental conditions in depositional environments as are volcanic ash beds, altered or unaltered.

Ordovician microfossils in general, and conodonts in particular, have proven to be a much better biostratigraphic tool than megafossils such as brachiopods, and the literature on the history of this problem is extensive (see Hall, 1986, for a summary). Nonetheless, even conodonts would appear to have their stratigraphic limitations, as shown in the above discussion, and a collaborative study using the Rocklandian K-bentonites as a stratigraphic datum around which the vertical distribution of conodonts can be shown would be a significant contribution to our understanding of Ordovician stratigraphy. Stratigraphic charts produced from such a study could be used to tie the vertical distribution of conodonts and other index fossils to the stratigraphic position of the Deicke and Millbrig K-bentonite Beds throughout the eastern midcontinent, but especially in the all-important southern Valley and Ridge, where Midcontinent and North Atlantic conodont faunas intermingle. The Deicke and Millbrig are in the *Phragmodus undatus* conodont zone throughout much of the eastern midcontinent (Huff and Kolata, 1990), and because both beds can now be identified at key exposures in the three confacies belts of Janusson and Bergström (1980) throughout the southern Valley and Ridge, it may be possible to match North Atlantic and Midcontinent conodont faunas with extreme precision. For example, because the Deicke and Millbrig have been identified at the Red Mountain Expressway, Tidwell Hollow, and Big Ridge sections, the vertical distribution of conodonts in those sections (Figs. 14, 17, and 18 of Hall, 1986) could now be tied directly to the location of K-bentonite beds in those sections (Figs. 27 and 31) with great precision, and probably with a minimum of resampling.

This discussion of just one particular area (central Alabama) shows the real need for collaboration of biostratigraphers and lithostratigraphers in preparing revised correlation charts for the Mohawkian based on the regional K-bentonite stratigraphic framework presented herein.

SUMMARY AND CONCLUSIONS

The discovery that the Deicke and Millbrig K-bentonite Beds can be reliably and consistently distinguished in the southeastern United States on the basis of their nonclay mineralogy has made it possible to overcome decades of confusion and frustration concerning the identification and correlation of K-bentonites at exposures of Rocklandian strata in that region. The importance of this to our understanding of Ordovician geology in this region cannot be overemphasized. A unified scheme of K-bentonite nomenclature is now a reality: beds T-3, V-3, B-3, R-7, and the Pencil Cave "metabentonite" are all the Deicke K-bentonite, and beds T-4, V-4, B-6, R-10, and the Mud Cave metabentonite are all the Millbrig, as shown in Figure 23. Regional variations in the diagenetic history of the Appalachian basin can be better understood following study of the K-bentonites in the different facies belts throughout the eastern midcontinent. It is hoped that with the recognition of correlatable K-bentonites in Rocklandian sections throughout the southern Appalachians, the K-bentonites will become a powerful tool in the investigation of biostratigraphic, geochemical, sedimentologic, and other geologic problems in the Rocklandian of this region, and further research into their geologic importance will be initiated. This could include, for example, a check of the biostratigraphic relations of the microfossils most widely used for correlation, i.e., conodonts and graptolites. Because the presence of recognizable time lines will improve the already important role of graptolites and especially conodonts as biostratigraphic markers in the Middle and Upper Ordovician of this region, there is great significance in having correlatable K-bentonites in measured sections on all thrust sheets where Rocklandian strata occur in the southern Valley and Ridge.

The major conclusions concerning the mineralogy, petrology, diagenetic history, and stratigraphic setting of the Deicke and Millbrig K-bentonites in the southeastern United States are:

1. The Deicke and Millbrig K-bentonite Beds are beds of altered volcanic ash, and this ash is entirely airfall in origin. Distinctly different nonclay mineral assemblages are present in the tuffaceous zones of each bed, and the two beds can be reliably and consistently distinguished on this basis. The major primary phenocrysts in the Deicke are labradorite and ilmenite; in the Millbrig they are andesine, quartz, and biotite. The Deicke contains trace amounts of quartz and biotite; both beds contain trace amounts of zircon and apatite. These distinct mineralogies indicate that the Deicke is altered dacitic or latitic ash, and the Millbrig is altered rhyodacitic ash.

2. The Deicke and Millbrig exhibit normal grading, but with notable differences. Whereas the Deicke is a single,

thick, graded bed, the Millbrig consists of as many as four separate graded sequences. These differences suggest that the Deicke is the product of a single, massive eruptive event, but that the Millbrig may be the product of several such events, or of one sustained event characterized by an evolving magma chamber. Both eruptions likely began with a Plinian phase, followed by a Peléan phase. The Plinian phase produced a shard-rich ash, and the Peléan phase a crystal-rich ash as a coignimbrite airfall deposit.

3. The volume (330 km^3 each postcompaction; 1,270 km^3 *minimum* each posteruption) and areal extent (estimated 600,000 km^2 *minimum* each) of both the Deicke and Millbrig are enormous. Comparison with the distribution of well-studied Cenozoic ashfalls suggests that the eruptions which produced the Deicke and Millbrig ashes were far greater, in every sense, than any eruption in recent history, including the Tambora eruption of 1815 and the Toba eruption of 75 Ka.

4. The primary plagioclases in the Deicke and Millbrig commonly are partly or completely replaced by authigenic feldspars. In Cincinnati Arch samples, only authigenic K-feldspar occurs, but both authigenic K-feldspar and authigenic albite occur in samples from the carbonate sequences of the western Valley and Ridge. This distribution indicates that the maximum post-Ordovician temperature encountered by the rocks along the Cincinnati Arch has never been more than about 70°C, whereas in the southern Valley and Ridge the maximum temperature was at a minimum 110°C, and in several areas probably exceeded 140°C.

5. The main clay mineral in the beds is mixed-layer illite/smectite. Kaolinite is present only in samples of the beds collected from the Colvin Mountain Sandstone and the Bays Formation in the eastern Valley and Ridge. The illite content of the I/S ranges from 70 to 80% along the Cincinnati Arch, 75 to 85% in the Valley and Ridge of Alabama and Georgia, and 80 to more than 90% in the Valley and Ridge of Virginia; the ordering of the I/S changes from R1 and R1/R3 along the Cincinnati Arch to predominantly R3 in the Valley and Ridge. These regional changes in the I/S reflect incremental increases in the maximum post-Ordovician temperatures experienced by the strata. It is evident that the ordering is a function of the thermal gradients established for the Ordovician of the southern Appalachian basin by Harris et al. (1978).

6. The migrating brine theory of Oliver (1986), together with the illitization theory of Morton (1985), adequately explains the distribution of authigenic feldspars in the Deicke and Millbrig, as well as regional variations in the composition of the I/S. The most likely source of the brines in the study area would appear to have been the shales and shaly lime mudstones in the Ordovician Sevier Basin and its equivalents (Kesler et al., 1988), which were dewatered during the Early and Middle Paleozoic. Dewatering of shales in older or younger basins at other times cannot be eliminated as a possible source of brines until absolute ages are obtained for the authigenic feldspars.

7. In the limestones of the Cincinnati Arch and the western Valley and Ridge province from Alabama to Virginia, ilmenite in the Deicke has been replaced by TiO_2 and pyrite. Minor pyrite also occurs in the Millbrig, although it is derived from the alteration of biotite, not ilmenite, and in most samples the biotites retain their original content of Ti, Fe, Mn, and K. In the quartz arenites of the Colvin Mountain Sandstone, the ilmenite in the Deicke is only slightly altered to leucoxene or it is unaltered. No pyrite, organic matter, primary feldspar, biotite, or even quartz (nondetrital) is present in samples of either the Deicke or Millbrig from the Colvin Mountain Sandstone. In the Moccasin Formation and the red Bays Formation, hematite is ubiquitous in the K-bentonites and in the surrounding sediments, but pyrite and ilmenite (in the Deicke) are absent, and biotite grains in the Millbrig are commonly extensively chloritized. In the green Bays Formation pyrite is present and hematite is absent except in a very few beds near the top of the Bays, and biotites in the Millbrig are little altered. These regional variations reflect the redox conditions of the pore waters during the period of diagenesis that had the most lasting effect on the K-bentonites and the adjacent sediments. In the limestones of the Cincinnati Arch and the western Valley and Ridge and in the clastics of the green Bays Formation, the major diagenetic changes occurred in the anoxic sulfidic environment of Berner (1981). In the Colvin Mountain Sandstone diagenesis occurred in the oxic environment of Berner (1981), and leaching was intense enough to remove all phenocrysts in both beds except the ilmenite in the Deicke. In the Moccasin and "red" Bays Formations, the most lasting episode of diagenesis occurred in the oxic environment, but the absence of pyrite and unaltered ilmenite in the Deicke indicates that the ilmenite had previously been altered in the anoxic, probably sulfidic environment. The iron released during the alteration of ilmenite was taken up by pyrite, which was later oxidized, possibly during the influx of meteoric waters from the tectonic highlands to the southeast. These meteoric waters evidently did not pass through the sediments of the green Bays, which, having never been oxidized, remain pyritic.

8. A general comparison of texture and composition suggests that the Millbrig is very likely the same K-bentonite as the "Big Bentonite" in the well-known Ordovician section at Kinnekulle, Sweden. Based on numerous additional criteria this suggested correlation has been formally proposed by Huff and others (1992).

9. The Deicke and Millbrig are the two thickest and most laterally persistent of the numerous Mohawkian K-bentonites of the eastern midcontinental U.S., as demonstrated by Kolata et al. (1986), Huff and Kolata (1990), Haynes (1992), and the present study. Because the base of the Deicke is a widely traceable horizon, and the Blackriveran-Rocklandian boundary can be placed at this horizon in most of the region, some "classic" Blackriveran strata, such as the upper Tyrone Limestone of Kentucky, the upper "Black River Limestone" of Indiana and Michigan, and the upper Carters Limestone of Tennessee and

Georgia, are in reality Rocklandian in age. This confirms the speculation of Huffman (1945), Gutstadt (1958), and others concerning the likely post-Blackriveran age of the above units.

10. The Deicke and Millbrig can be correlated between the Cincinnati Arch and the southern Valley and Ridge province, and along and across strike in the Valley and Ridge between Birmingham, Alabama, and Roanoke, Virginia. The Hagan section in Lee County, Virginia, is the key exposure for demonstrating the correlation from the Cincinnati Arch to the western Valley and Ridge, and for correlating the beds across strike in the Valley and Ridge. Both beds persist into the Valley and Ridge, with the Millbrig extending into the Bays Formation of northeast Tennessee and Virginia, where the Deicke is absent, and also into the Chickamauga Limestone of the Birmingham area, where the Deicke is thin to absent. At some exposures along the Cincinnati Arch the Millbrig is absent because of erosion associated with the post-Tyrone unconformity, and only the Deicke is present in some eastern exposures in Alabama, Georgia, and southeast Tennessee. Overall, the Millbrig is the most widespread and persistent of the two beds. These correlations have resulted in a significant reinterpretation of the K-bentonite stratigraphy in southwest Virginia, northeast Tennessee, and southeast West Virginia. The Deicke is bentonite V-3 of Rosenkrans (1936), not V-4, and the Millbrig is bentonite V-4 of Rosenkrans, not V-7. Bentonite V-7 of Rosenkrans is a separate bed that can be widely traced throughout much of southwest Virginia (Haynes, 1992).

11. The Deicke and Millbrig are useful in delineating the lateral extent of the unconformities that will be the underpinning of a sequence stratigraphic framework for the Middle and Upper Ordovician of this region. Using the K-bentonites as marker beds, it is evident that the unconformities with which they are most closely associated, the post-Tyrone, post-Chickamauga, and sub–Walker Mountain unconformities, are regionally restricted to varying extents.

12. The potential of the K-bentonites as marker beds is realized to a greater extent than ever before, which will make possible continual improvements in the conodont and graptolite biostratigraphic framework of the eastern midcontinent. Collaborative efforts between biostratigraphers and physical stratigraphers should eventually result in the development of correlation charts that tie in the vertical range of conodont species with the location of the K-bentonite beds in individual measured sections. This will for the first time allow a direct comparison of conodont forms from the Midcontinent and North Atlantic provinces, and it should be possible to establish unambiguous links between the two faunal assemblages.

13. The names "Deicke K-bentonite Bed" and "Millbrig K-bentonite Bed" which, as defined by Willman and Kolata (1978), are consistent with the North American Stratigraphic Code, are suitable for formal usage in the southeastern United States. In formal usage these names replace the informal alpha-numeric names V-3, B-3, T-3, and R-7 (Deicke), and V-4, B-6, T-4, and R-10 (Millbrig), of Rosenkrans (1936), Fox and Grant (1944), Wilson (1949), and Miller and Fuller (1954), respectively.

ACKNOWLEDGMENTS

This publication reports on the expansion and further distillation of a doctoral research project at the University of Cincinnati that was carried out under the direction of Warren D. Huff, whose guidance and encouragement since I began studying Ordovician K-bentonites have been invaluable. Initial field work was supported by National Science Foundation Grant EAR-8407018 (to W.D.H.). Richard L. Hay, University of Illinois–Urbana, a member of my dissertation committee, generously assisted with the study of the mineralogy and diagenetic history of the beds; his advice in the interpretation of thin sections was most helpful, and he also arranged for the use of the university's electron microprobe. The contributions of Profs. Huff and Hay to this study are particularly acknowledged. The advice given by Dennis R. Kolata, Illinois State Geological Survey, in the field is acknowledged, as is the field assistance of John S. Davis and Jeffrey L. Watson, two colleagues who assisted me in the all-important southwestern Virginia region in July, 1985, February, 1987, July, 1988, and August 1993. Critical comments in the field by Richard J. Diecchio, George Mason University, in June 1992, led to improvements in the discussion of stratigraphy. Eugene K. Rader, Virginia Division of Mineral Resources, helped me to understand many of the stratigraphic and structural complexities of the Ordovician in the Virginia Valley and Ridge province, and his important input over the years into this ongoing research project is acknowledged. The assistance and cheerful patience of Virginia Edwards, thin-section technician at the University of Cincinnati, is gratefully acknowledged, as is the draftsmanship of Lyla Messick, also at Cincinnati. Critical comments by reviewers John Dennison, Greg Ludvigson, Robert Sloan, and Brian Witzke greatly improved this manuscript. Finally, I also thank all the landowners and quarrymen, too numerous to mention individually, who provided access to the exposures I examined that are on private lands.

APPENDIX 1: SAMPLE LOCALITY INFORMATION AND ANALYTICAL METHODS

Part A. Sample localities

The following locality register is for the exposures and cores from which K-bentonite samples were collected and studied. The name of the locality, the appropriate U.S. Geological Survey 7½′ quadrangle, a brief description of the location, the citations of references with the best descriptions of the locality (if any), and the identifying numbers of the Deicke, Millbrig, and V-7 K-bentonite samples that were collected and studied, are given. Some additional exposures at which no K-bentonites were collected are included for their stratigraphic importance. For continuity in the geologic literature, locality names used are those of previous authors wherever possible.

(1) Hagan—Rose Hill 7½' Quadrangle., Lee County, Virginia. The Hardy Creek Limestone and Eggleston and Trenton Formations in exposure on the east side of the CSX (ex-L&N) rail switchback along the railroad right-of-way near State Road 621 about 0.8 km north of the intersection with U.S. Highway 58 (Huffman, 1945; Miller and Fuller, 1954, Miller and Brosgé 1954; Kreisa, 1980; Hall, 1986; Haynes, 1992).

Deicke: VA 2-1
Millbrig: VA 4-1

(2) Gate City–Gate City 7½' Quadrangle, Scott County, Virginia. The upper Moccasin, Eggleston, and lower Trenton (Martinsburg) Formations in exposure at the west end of Gateway Plaza, on the south side of combined U.S. Highways 23, 58, and 421 immediately before the gap through Clinch Mountain used by the highway and the Southern Railway (Haynes, 1992).

Deicke: VA:SC 1-1
Millbrig: VA:SC 1-2
V-7: VA:SC 1-3

(3) Rosedale—Elk Garden 7½' Quadrangle, Russell County, Virginia. The upper Moccasin, Eggleston, and lower Trenton (Martinsburg) Formations in exposure along the northeast side of old (i.e., up the embankment from new) State Highway 80, 2.3 km northwest of intersection with U.S. Highway 19 at Rosedale (Butts, 1940; Kreisa, 1980; Haynes, 1992).

Deicke: VA:RL 2-1
Millbrig: VA:RL 2-2
V-7: VA:RL 2-3

(4) Plum Creek—Tazewell South 7½' Quadrangle, Tazewell County, Virginia. Composite section. The upper Moccasin, Eggleston, and lower Trenton (Martinsburg) Formations in several exposures along the southeast side of State Highway 16 between Frog Level and Thompson Valley, along Plum Creek (Rosenkrans, 1936; Hergenroder, 1966; Haynes, 1992).

Deicke: VA:TZ 1-7
Millbrig: VA:TZ 1-8
V-7: VA:TZ 1-1

(5) Cove Creek—Cove Creek 7½' Quadrangle, Tazewell County, Virginia. The upper Moccasin and lower Eggleston Formations in exposure along State Road 614 at the first hairpin turn in road below Crabtree Gap (Haynes, 1992).

Deicke: VA:TZ 2-1
Millbrig: VA:TZ 2-2
V-7: VA:TZ 2-3

(6) Rocky Gap—Rocky Gap 7½' Quadrangle, Bland County, Virginia. The upper Moccasin and lower Eggleston Formations in exposure along the southwest side of U.S. Highway 52 (Frontage Road for Interstate Highway 77) 1.5 km south of Rocky Gap (Haynes, 1992).

Deicke: VA:BL 1-1
Millbrig: VA:BL 1-2

(7) Goodwins Ferry—Eggleston 7½' Quadrangle, Giles County, Virginia. The upper Moccasin Formation, Walker Mountain Sandstone Member, and Eggleston and lower Trenton (Martinsburg) Formations in exposure along the southeast side of State Road 625, on the east bank of the New River 500 m north (uphill) from the Norfolk & Western (ex-VGN) Railway grade crossing (Hergenroder, 1966; Kreisa, 1980; Haynes, 1992; Haynes and Goggin, 1993).

Deicke: VA:GI 4-0
V-7: VA:GI 4-1

(8) Mountain Lake Turnoff—Eggleston 7½' Quadrangle, Giles County, Virginia. The upper Witten Formation, Moccasin Formation, the Walker Mountain Sandstone Member, and the Eggleston Formation in exposure along the northeast side of westbound U.S. Highway 460, 400 m east of the intersection with State Road 700. The Moccasin and Walker Mountain are repeated by faulting just downsection of the gray Witten limestones (Hergenroder, 1966; Kreisa, 1980; Haynes, 1992; Haynes and Goggin, 1993).

Deicke: VA:GI 2-0

(9) Terrill Creek—Stony Point 7½' Quadrangle, Hawkins County, Tennessee. The Walker Mountain Sandstone Member, upper Bays Formation, and lower Trenton (Martinsburg) Formation in interrupted exposures along the northeast side of the road along Terrill Creek, in the gap between Hennard and River Mountains 1.3 km due south of Miller Island in the Clinch River (Hergenroder, 1966; Haynes, 1992)

Millbrig: TN:HK 2-1

(10) Chatham Hill—Chatham Hill 7½' Quadrangle, Smyth County, Virginia. The Walker Mountain Sandstone Member and the upper Bays Formation in exposure along the northeast side of State Highway 16 on the northwest slope of Big Walker Mountain 3.7 km south of Chatham Hill (Hergenroder, 1966; Kreisa, 1980; Haynes, 1992; Haynes and Goggin, 1993).

Millbrig: VA:SM 1-3 through 1-6

11) Big Walker Lookout—Big Bend 7½' Quadrangle, Bland County, Virginia. The Bays Formation along the south side of U.S. Highway 52, at the first hairpin turn on the mountain going south toward the lookout tower atop Big Walker Mountain (Haynes, 1989).

Millbrig: VA:BL 2-1

(12) Crockett Cove—Wytheville 7½' Quadrangle, Wythe County, Virginia. The Witten Formation, Walker Mountain Sandstone Member, and lower Bays Formation in exposure along the northeast side of northbound Interstate Highway 77, 2 km northwest of the I-77/I-81 interchange at Wytheville (Haynes, 1992; Haynes and Goggin, 1993).

Millbrig: VA:WY 2-1

(13) Connor Valley—Fosters Falls 7½' Quadrangle, Wythe County, Virginia. Two separate exposures for a composite section. Connor Valley North: the upper Wassum Formation, Walker Mountain Sandstone Member, and lower Bays Formation in exposure along Frontage Road for Interstate Highway 81 just south of T intersection with State Road 100, which is 500 m north of the cloverleaf interchange of I-81 and State Road 100. Connor Valley South: the middle and upper Bays are exposed along the east side of State Road 726 in Connor Valley at the west end of a small hill that the road and creek skirt to the south 1.3 km north-northeast of Crockett Knob (Hergenroder, 1966; Haynes, 1992; Haynes and Goggin, 1993).

Millbrig: VA:WY 1-1
V-7: VA:WY 1-2

(14) Gap Mountain—Radford North 7½' Quadrangle, Giles County, Virginia. The Witten and Moccasin Formations, Walker Mountain Sandstone Member, and Eggleston and Trenton (Martinsburg) Formations in exposure along the northeast side of the Norfolk & Western (ex-VGN) Railway right-of-way parallel to State Road 625 along the east side of the New River about 500 m north of Big Falls (Tuscarora Sandstone outcrop) in the New River. The Walker Mountain is exposed in the ravine up the hill from the culvert, but the Deicke is covered (Hergenroder, 1966; Kreisa, 1980; Haynes, 1992; Haynes and Goggin, 1993).

Millbrig: VA:GI 1-1
V-7: VA:GI 1-2

(15) Catawba—Catawba 7½' Quadrangle, Roanoke County, Virginia. The upper Bays and lower Trenton (Martinsburg) Formations in exposure along the southwest side of State Highway 311, 50 m southeast of the bridge over Catawba Creek (Hergenroder, 1966; Kreisa, 1980; Haynes, 1992).

Millbrig: VA:RO 1-4
V-7: VA:RO 1-3

(16) Daleville—Daleville 7½' Quadrangle, Botetourt County,

Virginia. The upper Liberty Hall Formation, Walker Mountain Sandstone Member, Bays Formation, and lower Trenton (Martinsburg) Formations in exposure along the east side of the right-of-way of the Norfolk & Western Railway's Lone Star spur 800 m due west of Lord Botetourt High School (Hergenroder, 1966; Haynes, 1992).

Millbrig: VA:BT 2-1
V-7: VA:BT 2-2

(17) Harrogate—Middlesboro South 7½′ Quadrangle, Claiborne County, Tennessee. The upper Eggleston and lower Trenton (Martinsburg) Formations in exposure along the southwest side of the dismantled CSX (ex-L&N) rail right-of-way about 500 m southeast of the former grade crossing at Myers School (Huffman, 1945; Haynes, 1992).

Deicke: TN:CL 1-1
Millbrig: TN:CL 1-2

(18) Hinds Creek Quarry—Powell 7½′ Quadrangle, Anderson County, Tennessee. The upper Chickamauga Limestone (upper Carters and Eggleston equivalent) exposed in the inactive quarry on Brushy Mountain Road in the Lone Mountain subdivision along Hinds Creek (Coker, 1962; Haynes, 1992).

Deicke: TN 10
Millbrig: TN 11-1 through 11-4

(19) Henson Creek—Henson Gap 7½′ Quadrangle, Sequatchie County, Tennessee. The Upper Carters Limestone and lower Hermitage Formation exposed along the southeast side of Henson Creek near Davis Chapel Cemetery (Milici, 1969).

Deicke: TN:SQ 1-1
Millbrig: TN:SQ 1-2

(20) Tillet Brothers Quarry—Normandy 7½′ Quadrangle, Bedford County, Tennessee. The Carters Limestone and lower Hermitage Formation exposed in the Tillet Brothers Paving Company Quarry (the "Stone Man") on U.S. Highway 41 Alternate, 8 km southeast of Shelbyville (Haynes, 1989).

Deicke: TN 6
Millbrig: TN 7-1 through 7-6

(21) South Carthage—Gordonsville 7½′ Quadrangle, Smith County, Tennessee. The Upper Carters Limestone and lower Hermitage Formation in exposures on the west side of State Highway 53 about 3.7 km south of the intersection with U.S. Highway 70N, and along the north side of U.S. Highway 70N at that intersection (Haynes, 1989; Huff and Kolata, 1990).

Deicke: TN 1-1 through 1-3
Millbrig: TN 2 and TN 4

(22) Davis Crossroads—Kensington 7½′ Quadrangle, Walker County, Georgia. The Carters Limestone and Hermitage Formation along the dismantled Southern (ex-TA&G) Railway right-of-way about 800 m southeast of the grade crossing with State Road 341 at the Barwick Mill (Allen and Lester, 1957; Milici and Smith, 1969).

Deicke: GA:WL 3-1
Millbrig: GA:WL 3-2

(23) Rising Fawn—Sulphur Springs 7½′ Quadrangle, Dade County, Georgia. The Upper Carters Limestone and lower Hermitage Formation along the west side of the right-of-way of the Southern Railway (Alabama Great Southern) about 2 km south of Rising Fawn, in the second cut from the north (Haynes, 1989).

Deicke: GA:DD 2-1
Millbrig: GA:DD 2-2

(24) Fort Payne—Fort Payne 7½′ Quadrangle, Dekalb County, Alabama. The Stones River Group in exposure along the southeast side of northbound Interstate Highway 59 just north of Interchange 222 (Drahovzal and Neathery, 1971; Benson and Stock, 1986; Huff and Kolata, 1990).

Deicke: AL 12-1 through 12-4
Millbrig: AL 14-1 and 14-2

(25) Big Ridge—Keener 7½′ Quadrangle, Etowah County, Alabama. The Stones River Group in exposure along the northeast side of northbound Interstate Highway 59 in the cut through Big Ridge about 22 km northeast of the Gadsden-Attalla exit (Drahovzal and Neathery, 1971; Chowns and McKinney, 1980; Kunk and Sutter, 1984; Hall, 1986).

Deicke: AL 7-1 through 7-5
Millbrig: AL 9-1 and 9-2

(26) Bruton Gap—Gadsden 7½′ Quadrangle, Etowah County, Alabama. The Stones River Group in exposure along the northeast side of State Road 227 about 150 m west of the crest over Big Ridge in Bruton Gap (Ward, 1983).

Deicke: AL:ET 2-1
Millbrig: AL:ET 2-6

(27) Tidwell Hollow—Cleveland 7½′ Quadrangle, Blount County, Alabama. The Chickamauga Limestone undivided in exposure along the northwest side of State Highway 15 just west of Tidwell Hollow about 5 km southeast of Locust Fork (Benson and Stock, 1986; Hall, 1986).

Millbrig: AL:BL 1-1

(28) Alexander Gap—Glencoe 7½′ Quadrangle, Calhoun County, Alabama. The upper Greensport Formation and the Colvin Mountain Sandstone in exposure along the east side of northbound U.S. Highway 431 in Alexander Gap just south of the initial downgrade through the gap (Jenkins, 1984).

Deicke: AL:CH 1-2
Millbrig: AL:CH 1-4

(29) Red Mountain Expressway—Birmingham South 7½′ Quadrangle, Jefferson County, Alabama. The Chickamauga Limestone undivided in exposures on both sides of U.S. Highway 31, the Red Mountain Expressway, in the deep cut through Red Mountain (Drahovzal and Neathery, 1971; Raymond, 1978; Chowns and McKinney, 1980; Hall, 1986; Benson and Stock, 1986).

Deicke: AL 1
Millbrig: AL 3-1 through 3-7

(30) Green Springs—Birmingham South 7½′ Quadrangle, Jefferson County, Alabama. The Chickamauga Limestone undivided in the exposure along the west side of southbound Interstate Highway 65 just north of the Green Springs exit (Lee, 1983).

Millbrig: AL 4

(31) Old North Ragland Quarry—Ragland 7½′ Quadrangle, St. Clair County, Alabama. The Little Oak Limestone in an abandoned quarry on the southwest end of low ridge just north of the mouth of Trout Creek along the Coosa River. This section is included because of the possibility that the thick biotitic K-bentonite is the Millbrig. This K-bentonite is enigmatic; conodonts from the surrounding Little Oak Limestone at this quarry suggest a Chazyan to lowermost Blackriveran age (Drahovzal and Neathery, 1971; Schmidt, 1982), but using the $^{40}Ar/^{39}Ar$ method, M. J. Kunk, U.S. Geological Survey (personal communication), has obtained uppermost Blackriveran to Rocklandian (~454-455 Ma) ages from biotites in this K-bentonite.

Millbrig?: AL:SC 1-1

(32) High Bridge—Wilmore 7½′ Quadrangle, Jessamine County, Kentucky. The upper Tyrone and lower Lexington Limestones exposed in the quarry adjacent to and in low outcrops along the right-of-way of the Southern (CNO&TP) Railway. The quarry is about 400 m north of the bridge, and the rail cuts are about 1.3 km north of the bridge (Nelson, 1922; Huffman, 1945; Kerr and Kulp, 1949; Cressman and Noger, 1976; Scharpf, 1990; Haynes, 1992).

Deicke: KRI 4-0
Millbrig: KRI 5-0

(33) Lexington Limestone Quarry—Nicholasville 7½′ Quadrangle, Jessamine County, Kentucky. The upper Tyrone and lower

Lexington Limestones exposed in the adit to the underground workings at the quarry, which is west of Nicholasville on Catnip Hill Road about 1.5 km west of the intersection with U.S. Highway 27; the Millbrig is absent by erosion (Haynes, 1989; Scharpf, 1990).

Deicke: KY 6-1 through 6-6

(34) Black River Mine—Moscow 7½′ Quadrangle, Pendleton County, Kentucky. The upper Tyrone and lower Lexington Limestones exposed in the shaft to the underground workings at the Black River Mine, Carntown (Kunk and Sutter, 1984; Haynes, 1992).

Deicke: BRM T-3
Millbrig: BRM T-4

(35) Core samples—Patriot 7½′ Quadrangle, Gallatin County, Kentucky. Tyrone Limestone in core from Dravo Corp., BB-77-5, Carter Coordinates 24-CC-57 (from the collection of Warren Huff, University of Cincinnati) (Huff, 1963; 1983b).

Deicke: BB 5-5-1
Millbrig: BB 5-2-3

(36) Fincastle—Oriskany 7½′ Quadrangle, Botetourt County, Virginia. The Paperville Shale and the Martinsburg Formation, including the Fincastle Conglomerate Member, exposed in cuts along U.S. Highway 220 and adjacent quarry (now an equipment storage area) just north of Fincastle, Virginia. The Fincastle is considered equivalent to the Bays Formation and the Walker Mountain Sandstone Member (Haynes, 1992, Fig. 26); no K-bentonites were found at any exposure (Rader and Gathright, 1986; Zhenzhong and Eriksson, 1991).

(37) Horseleg Mountain—Rome South 7½′ Quadrangle, Floyd County, Georgia. The Greensport Formation and Colvin Mountain Sandstone exposed at the S-curve along Mt. Alto Road atop Horseleg Mountain (Chowns and Carter, 1983b).

Deicke: GA:FL 1-1

(38) Core samples—Madeira 7½′ Quadrangle, Hamilton County, Ohio. Tyrone Limestone in core from Newtown, Dravo Corp., K.C. Bowman #1 (N-80), SE¼NW¼Sec.2, T4N R2W (from the collection of Warren Huff, University of Cincinnati) (Huff, 1983b).

Deicke: N-5
Millbrig: N-2

(39) Taylor Avenue Quarry—Frankfort West 7½′ Quadrangle, Franklin County, Kentucky. The upper Tyrone and lower Lexington Limestones in quarry on State Road 1211 (Taylor Avenue) just south of CSX (ex-L&N) railroad bridge over the Kentucky River in Frankfort and about 0.8 km south of the intersection with Benson Valley Road; the Millbrig is absent by erosion (Juniper Hill section of Scharpf, 1990).

Deicke: KY 10-1 through 10-4

(40) Shakertown—Harrodsburg 7½′ Quadrangle, Mercer County, Kentucky. The upper Tyrone and lower Lexington Limestones exposed along the east side of U.S. Highway 68 about 1.5 km southwest of the intersection with State Road 33 at Shakertown (Scharpf, 1990).

Millbrig: KY:ME 1-1

(41) Core samples—Gravel Switch 7½′ Quadrangle, Marion County, Kentucky. Tyrone and Lexington Limestones in core from Humble Oil & Refining Co., B. Lanham #1, Carter Coordinates 17-N-54 (from the collection of Warren Huff, University of Cincinnati).

Deicke: C-114 T-3
Millbrig: C-114 T-4

(42) Core samples—Yosemite 7½′ Quadrangle, Casey County, Kentucky. Tyrone and Lexington Limestones in core from Humble Oil & Refining Co., W.C. Patterson #1, Carter Coordinates 14-K-57 (from the collection of Warren Huff, University of Cincinnati).

Deicke: C-113 T-3
Millbrig: C-113 T-4

(43) Core samples—Eli 7½′ Quadrangle, Russell County, Kentucky. Tyrone and Lexington Limestones in core from Cominco American, Omer Roy #1, Carter Coordinates 10-G-55 (from the collection of Warren Huff, University of Cincinnati).

Millbrig: C-211

(44) Core samples—Fountain Run 7½′ Quadrangle, Monroe County, Kentucky. Tyrone and Lexington Limestones in core from St. Joe Minerals Corp., Hole 78K-1, Carter Coordinates C-43 (from the collection of Warren Huff, University of Cincinnati).

Deicke: KY 1-3
Millbrig: KY 1-4

(45) Core samples—Smithville 7½′ Quadrangle, Clay County, Tennessee. Carters Limestone in core from St. Joe Minerals Corp., Hole 78T-5 (#T-DK-1), Carter Coordinates 7S-46E (from the collection of Warren Huff, University of Cincinnati).

Deicke: TN 3-3
Millbrig: TN 3-8

(46) Core samples—Center Hill Dam 7½′ Quadrangle, DeKalb County, Tennessee. Carters Limestone in core from U.S. Army Corps of Engineers, Hole #DC-100 Center Hill Dam, Carter Coordinates 5-47E-9S (from the collection of Warren Huff, University of Cincinnati).

Deicke: CH T-3

(47) Core samples—Columbus City 7½′ Quadrangle, Marshall County, Alabama. Carters Limestone in core from Tennessee Valley Authority, Hole #AH-40 Murphy Hill, NE¼Sec.25,T6S R4E (from the collection of Warren Huff, University of Cincinnati) (Günal, 1979).

Deicke: KB-35
Millbrig: KB-34

(48) Greensport Gap—Ohatchee 7½′ Quadrangle, Etowah County, Alabama. The upper Greensport Formation and the Colvin Mountain Sandstone in exposure along the east side of State Road 77 in the cut through Beaver Creek Mountain (Drahovzal and Neathery, 1971; Jenkins, 1984).

Deicke: AL:ET 1-2

(49) Dirtseller Mountain—Gaylesville 7½′ Quadrangle, Cherokee County, Alabama. Stones River Group exposed in small inactive quarry off of State Road 75 (Drahovzal and Neathery, 1971).

Millbrig: AL 17

(50) Trenton—Trenton 7½′ Quadrangle, Dade County, Georgia. Lower beds of the Upper Carters Limestone exposed along the northwest side of the entrance ramp to southbound Interstate Highway 59 from State Road 136 at bridge over I-59 just southwest of downtown Trenton (Haynes, 1989).

Deicke: GA:DD 1-1

(51) Mill Creek—Kensington 7½′ Quadrangle, Walker County, Georgia. Lower beds of the Upper Carters Limestone in exposure along the east side of State Road 193 about 100 m north of the bridge over Mill Creek. Nearby section along Mill Creek is described by Milici and Smith (1969).

Deicke: GA:WL 1-1

(52) Old Bethel Church—Kensington 7½′ Quadrangle, Walker County, Georgia. Upper Carters Limestone and lower Hermitage Formation in the small borrow pit along McLemore Cove Road just northeast of Old Bethel Church, directly beneath the powerline right-of-way and across the road from the electrical substation. The section continues for some distance along the right-of-way. By virtue of their location along an anticlinal axis the strata are flat-lying and appear essentially undisrupted structurally; by contrast, the K-bentonite is visibly contorted between its lower and upper contacts, evidently from having been the horizon along which a zone of bedding parallel slip occurred (Milici and Smith, 1969; Hall, 1986).

Deicke: GA:WL 2-1

(53) Dodds Avenue quarry—Chattanooga 7½′ Quadrangle, Hamilton County, Tennessee. The lower beds of the Upper Carters Limestone exposed in the small borrow pit along East 28th Street at intersection with Dodds Avenue, in Chattanooga (Haynes, 1989).

Deicke: TN:HM 1-1

(54) Windsor Street quarry—Chattanooga 7½′ Quadrangle, Hamilton County, Tennessee. The lower beds of the Upper Carters Limestone exposed along the east side of the abandoned quarry, now a childrens' playground, on Windsor Street just east of the intersection with Dodson Road, in Chattanooga (Haynes, 1989).

Deicke: TN:HM 2-1

(55) Pikeville (Stephens Road)—Melvine 7½′ Quadrangle, Bledsoe County, Tennessee. The Carters Limestone exposed in roadcut along Stephens Road about 2.2 km from intersection with U.S. Highway 127 near Litton (Haynes, 1989).

Deicke: TN 12-1 through 12-3

(56) Eidson—Kyles Ford 7½′ Quadrangle, Hawkins County, Tennessee. The Moccasin and lower Eggleston Formations exposed along the south side of State Highway 70 between Eidson and Rogersville near the base of Clinch Mountain on the northwest slope below Little War Gap (Hergenroder, 1966; Haynes, 1992).

Deicke: TN:HK 3-1
Millbrig: TN:HK 3-2

(57) Hurricane Bridge—Hubbard Springs 7½′ Quadrangle, Lee County, Virginia. The upper Hardy Creek Limestone and lower Eggleston Formation in exposure along the north side of State Road 654 southeast of Hurricane Bridge over the Powell River (Miller and Brosgé, 1954; Haynes, 1992).

Deicke: VA:LE 1-1
Millbrig: VA:LE 1-2

(58) Dryden—Keokee 7½′ Quadrangle, Lee County, Virginia. The upper Eggleston Formation and the lower Trenton Formation exposed along the west side of the CSX (ex-L&N) railroad right-of-way at the big bend in the Powell River near Dryden (Miller and Englund, 1975).

Millbrig: VA:LE 3-1

(59) Bluefield—Bastian 7½′ Quadrangle, Mercer County, West Virginia. The Moccasin and lower Eggleston Formations exposed along the south side of State Road 598 (formerly U.S. Highway 52) on the northwest slope of East River Mountain about 2.5 km below the mountain crest (Woodward, 1951).

Deicke: WV:ME 1-1
Millbrig: WV:ME 1-2

(60) The Narrows—Narrows 7½′ Quadrangle, Giles County, Virginia. The upper Eggleston Formation exposed along the north side of westbound U.S. Highway 460 in The Narrows of the New River (Rosenkrans, 1936; Woodward, 1951; Kay, 1956; Hergenroder, 1966; Kreisa, 1980).

V-7: VA:GI 3-1

(61) Trigg—Staffordsville 7½′ Quadrangle, Giles County, Virginia. The upper Moccasin Formation, Walker Mountain Sandstone Member, and Eggleston and lower Trenton (Martinsburg) Formations in exposure along State Road 730 just west of the intersection with State Road 622 (Hergenroder, 1966; Haynes and Goggin, 1993).

Deicke: VA:GI 5-1
Millbrig?: VA:GI 5-2
V-7: VA:GI 5-3

(62) Thorn Hill—Avondale 7½′ Quadrangle, Grainger County, Tennessee. The Moccasin Formation in exposures along U.S. Highway 25E on northwest slope of Clinch Mountain (Simonson, 1985).

Deicke: TN 14-1 through 14-4
Millbrig: TN:GR 1-2

(63) Nebo—Nebo 7½′ Quadrangle, Smyth County, Virginia. The Walker Mountain Sandstone Member and adjacent strata of the red Bays Formation in exposure on the east side of State Road 622 on the northwest side of Big Walker Mountain 2 km south of Nebo; no K-bentonites were found at this exposure (Handwerk, 1981).

(64) Ellett—Ironto 7½′ Quadrangle, Montgomery County, Virginia. The upper Liberty Hall Formation, Walker Mountain Sandstone Member, and Bays and lower Trenton (Martinsburg) Formations in several exposures along State Roads 603 and 641 beginning just east of the bridge over Wilson Creek. The several K-bentonites in the structurally disrupted upper Bays at this exposure are interpreted as being equivalent to beds V-7 or higher of Rosenkrans (1936), not the Deicke or Millbrig (Butts, 1940; Hergenroder, 1966; Haynes, 1992; Haynes and Goggin, 1993).

(65) Peters Creek—Salem 7½′ Quadrangle, Roanoke County, Virginia. The upper Liberty Hall Formation, Walker Mountain Sandstone Member, and lower Bays Formation in low bluff along State Road 629 at the headwaters of Peters Creek about 2.1 km east of Hanging Rock; no K-bentonites were found at this exposure (Haynes, 1992; Haynes and Goggin, 1993).

(66) Cliff Dale Chapel South—Jordan Mines 7½′ Quadrangle, Alleghany County, Virginia. The McGlone Limestone (Edinburg of Lesure, 1957), Walker Mountain Sandstone Member, and Eggleston and Dolly Ridge Formations exposed along the north side of State Road 617, 1.5 km southeast of old Cliff Dale Chapel. Following Lesure (1957), the Eggleston here includes both the Moccasin and the Eggleston Formations of Kay's 1956 sections from the Rich Patch Valley (Kay, 1956; Lesure, 1957; Haynes and Goggin, 1993).

Deicke?: VA:AL 1-1
Millbrig: VA:AL 1-2
V-7: VA:AL 1-3

(67) Cove Spring Road—Frankfort East 7½′ Quadrangle, Franklin County, Kentucky. The upper Tyrone and lower Lexington Limestones exposed along Cove Spring Road and U.S. Highway 127 north. Here the Millbrig is absent by erosion and the Deicke is much thinner (21 cm) than at the many exposures farther east such as High Bridge and the Lexington Limestone Quarry, a thickness trend that is shown in Figure 17 of Cressman and Noger (1976).

Deicke: KY:FK 1-1

(68) Citico Beach—Vonore 7½′ Quadrangle, Monroe County, Tennessee. The upper Bays Formation (Ordovician) and the Chattanooga Shale (Devonian-Mississippian) on the road to Citico Beach 400 m east of the access road to the Tanasi-Chota Memorials on Tellico Lake. The Deicke is the uppermost bed of the Bays, and it overlies red shale of the Bays and underlies the dark brown shale of the Chattanooga (Rodgers, 1953).

Deicke: TN:MR 1-1

(69) Millers Cove—Glenvar 7½′ Quadrangle, Roanoke County, Virginia. The upper Liberty Hall Formation, Walker Mountain Sandstone Member, and lower Bays Formation in field immediately northeast of the intersection of State Roads 701 and 620, and upper Bays Formation along State Road 701 about 350 m southeast of that intersection, for a composite section. At this exposure the Walker Mountain is a limestone pebble conglomerate that is strikingly similar to some beds in the Fincastle Conglomerate Member (Rosenkrans, 1936; Kay, 1956; Hergenroder, 1966; Haynes and Goggin, 1993).

Millbrig?: VA:RO 2-1

Part B. Methods of sample collection, preparation, and analysis

In this study K-bentonite samples were collected and examined from cores and outcrops of Rocklandian strata throughout the southeastern United States. Field work was carried out during June, August, and December 1984; March and July, 1985; February and March,

1986; February, May, and July, 1987; July 1988; February, March, and June 1991; May, June, and August 1992; and March, April, and August 1993. In addition to the localities listed in Part A of this appendix, many other outcrops were visited but no samples were collected because of a lack of satisfactory exposure or because K-bentonites were not present. At all outcrops, extensive excavation was usually required because K-bentonites weather recessively, even in relatively new exposures. In general, float and surficial material usually had to be cleared away; the bed was then excavated with hand-held mattocks and shovels to expose relatively unweathered material. Where hand digging was inadequate to expose unweathered material, samples were collected as far back into the outcrop as possible. Samples were collected systematically from texturally different intervals (as identified with a hand lens), and the overall thickness of the K-bentonite, as well as the thickness of each visually distinct zone, was measured. Some phenocrysts could usually be identified with a hand lens, allowing for preliminary comparisons of each measured section. Previously described and measured sections were used whenever possible, with a check of the accuracy of the measurements. In most cases, every described K-bentonite bed or unit in a measured section had been assigned a number by the geologist who wrote the description, and thus any samples that I collected from a particular section could be tied in with the existing description for future reference. If no satisfactory measured section existed, the section was measured and described to place the location of each K-bentonite.

In addition to samples obtained from field work, 11 samples from the core collection of Warren D. Huff of the University of Cincinnati were studied. These samples came mostly from the Cumberland Saddle region of the Cincinnati Arch, with one sample from northern Alabama.

Thin-section analysis was the primary means of laboratory investigation, using the petrographic microscope and the electron microprobe. Some grain separates were also prepared. The analysis of clay minerals performed separately, by powder x-ray diffraction.

Because of their smectite content, thin sections of bentonites and K-bentonites are notoriously difficult to prepare, and thus some special preparation techniques were required. Most of the samples disintegrated in water when wetted; therefore, water was an unsatisfactory coolant and most thin sections had to be prepared dry and by hand grinding. This was accomplished by first drying the samples overnight at 60°C. After drying, a flat surface suitable for bonding to a petrographic slide was prepared by grinding the sample on medium- to coarse-grit emery cloth laid on a clean, flat glass plate. Well-indurated samples such as those from cores did not crumble during the initial grinding. This flat face was then ground smoother with successively finer grit emery cloth, using 320- and then 400-grit cloth. All grinding was done on the glass plate to maintain a flat surface. After the final grinding and rough polishing with 600 grit cloth, the flat surface was brushed clean of dust with a soft-bristle toothbrush, bonded to a frosted glass petrographic slide with Hillquist petrographic epoxy, and allowed to dry overnight at room temperature. This chip was trimmed off with a rotary trim saw using mineral oil or propanol as coolant, and the section was then manually ground down again, with successively finer grit emery cloth, to a thickness of 30μ. Fifty-eight thin sections were successfully prepared using this method.

The less-indurated samples crumbled after the first attempt at preparing a flat face with emery cloth. Impregnation with petrographic epoxy in a pressurized chamber was used in an attempt to solve the problem. This worked with only limited success, however. When it did work, the impregnated samples were first sawed on a trim saw with mineral oil as coolant, and the flat face of the sawed surface was ground using emery cloth, and then bonded, sawed off, ground down, and polished dry on emery cloth down to 600-grit paper, as described above. Twenty-four thin sections were successfully prepared using this method. After final brushing to remove dust, the thin section was ready for study with the petrographic microscope, where identification of nonclay minerals was made using standard optical methods.

Additional polishing for electron microprobe analysis was carried out with a polishing lap using 6.0-, 3.0-, and 1.0-μ diamond pastes with polishing oil as a coolant and lubricant. Of the 82 thin sections made, 25 were polished in this manner.

Some samples were too weathered or poorly indurated to make thin sections, even with pressure impregnation, so grain separates were prepared. Dampened samples were soaked in kerosene for an hour, after which the kerosene was poured off and water added. This immediately and effectively dispersed the clays, and the mixture was then wet-sieved through a 200-mesh screen, allowing the clays but not the larger phenocrysts to pass through. The grains remaining on the screen were placed in glass dishes, dried overnight at 60°C, and stored in labeled vials for later examination under the petrographic or binocular microscope.

X-ray diffraction (XRD) patterns were produced by x-ray analysis of oriented glycol-solvated mounts of the <2.0-μm size fraction of selected Deicke and Millbrig samples, two samples of K-bentonite V-7, and five other K-bentonites. To obtain the <2.0-μm size fraction, bulk K-bentonite samples were first dispersed in deionized water using an ultrasonic bath and a rotary mixer. This was followed by gravity settling to collect the <2.0-μm size fraction from suspension; after sitting for 230 min, the upper part of the liquid was poured off and centrifuged. The clay, which was now removed from suspension, was placed onto petrographic well slides with a small flat spatula using the smear method, thus producing a preferred orientation of the clay minerals.

Identification of mixed-layer illite/smectite is based on the interpretation of XRD patterns of ethylene glycol–solvated samples. To glycolate the samples, the glass slides were set on an elevated tray with legs that itself rested inside a glass bell jar having ethylene glycol pooled at the bottom. The glass lid was sealed with a silicone sealant, and the jar and its enclosed contents were placed in an oven overnight at 60°C. The jar was then removed from the oven and placed aside for 3 to 4 days, with the lid still sealed, until the samples were analyzed. X-ray diffraction was performed at the University of Cincinnati on a Siemens D-500 automated diffractometer using CuKα radiation. The scanning speed was 0.05°/2-sec interval, and the scans were from 2° to 60° 2Θ, with occasional scans rerun at slower speeds or between a smaller interval.

Most microprobe analyses were carried out at the University of Illinois on a JEOL two-channel electron microprobe, but three samples were analyzed at the University of Cincinnati on an ARL three-channel electron microprobe, both of which had 5.0-μ-wide beams. The microprobe was calibrated to an internal standard prior to the sample analyses. Polished thin sections were first coated with a carbon film, and particular grains that had been chosen for analysis by viewing with the petrographic microscope were suitably marked on the slide to facilitate locating them in the optical system of the electron microprobe. The slides were then affixed to a brass holder with a carbon or silver paste to ground the thin section to the machine, after which the paste was then allowed to dry. When dry, the slides were loaded into the sample chamber and the analyses begun. Standard data reduction methods were used to determine mineral composition from the analytical printout produced by each analysis.

REFERENCES CITED

Ali, A. D., and Turner, P., 1982, Authigenic K-feldspar in the Bromsgrove Sandstone Formation (Triassic) of central England: Journal of Sedimentary Petrology, v. 52, p. 187–197.

Allen, A. T., and Lester, J. G., 1957, Zonation of the Middle and Upper Ordovician strata in northwestern Georgia: Geological Survey of Georgia Bulletin 66, 110 p.

Altaner, S. P., Hower, J., Whitney, G., and Aronson, J. L., 1984, Model for K-bentonite formation: Evidence from zoned K-bentonites in the disturbed belt, Montana: Geology, v. 12, p. 412–415.

Bates, R. L., 1939, Geology of Powell Valley in northwestern Lee County, Virginia: Virginia Geological Survey Bulletin 51-B, p. 31–94.

Bay, H. X., and Munyan, A. C., 1940, The bleaching clays of Georgia, *in* Clay investigations in the southern states, 1934-35: U.S. Geological Survey Bulletin 901, p. 251–300.

Beall, A. O., Jr., and Ojakangas, R. W., 1967, Mineralogy of an Upper Cambrian K-bentonite from Missouri: Journal of Sedimentary Petrology, v. 37, p. 952–956.

Bearce, D .N., 1973, Origin of conglomerates in Silurian Red Mountain Formation of central Alabama: Their paleogeographic and tectonic significance: American Association of Petroleum Geologists Bulletin, v. 57, p. 688–701.

Benson, D. J., 1986, Depositional setting and history of the Middle Ordovician of the Alabama Appalachians, *in* Benson, D. J., and Stock, C. W., eds., Depositional history of the Middle Ordovician of the Alabama Appalachians (23rd Annual Field Trip Guidebook): Tuscaloosa, Alabama Geological Society, p. 15–31.

Benson, D. J., and Stock, C. W., eds., 1986, Depositional history of the Middle Ordovician of the Alabama Appalachians: 23rd Annual Field Trip Guidebook, Tuscaloosa, Alabama Geological Society, 120 p.

Berner, R. A., 1981, A new geochemical classification of sedimentary environments: Journal of Sedimentary Petrology, v. 51, p. 359–365.

Best, M. G., Christiansen, E. H., and Blank, R. H., Jr., 1989, Oligocene caldera complex and calc-alkaline tuffs and lavas of the Indian Peak volcanic field, Nevada and Utah: Geological Society of America Bulletin, v. 101, p. 1076–1090.

Boles, J. R., 1982, Active albitization of plagioclase, Gulf Coast Tertiary: American Journal of Science, v. 282, p. 165–180.

Boles, J. R., and Franks, S. G., 1979, Clay diagenesis in Wilcox sandstones of southwest Texas: Implications of smectite diagenesis on sandstone cementation: Journal of Sedimentary Petrology, v. 49, p. 55–70.

Borchardt, G. A., Harward, M. E., and Schmitt, R. A., 1971, Correlation of volcanic ash deposits by activation analysis of glass separates: Quaternary Research, v. 1, p. 247–260.

Borchardt, G. A., Norgren, J. A., and Harward, M. E., 1973, Correlation of ash layers in peat bogs of eastern Oregon: Geological Society of America Bulletin, v. 84, p. 3101–3108.

Bowen, R. L., 1967, Volcanic events of the Middle Ordovician in eastern North America: 48th Annual Meeting, Transactions, American Geophysical Union, v. 48, p. 226.

Brockman, G. F., 1978, Pre-Pennsylvanian stratigraphy of the Birmingham area, *in* Kidd, J. T., and Shannon, S. W., eds., Stratigraphy and structure of the Birmingham area, Jefferson County, Alabama (16th Annual Field Trip Guidebook): Tuscaloosa, Alabama Geological Society, p. 1–14.

Brusewitz, A. M., 1986, Chemical and physical properties of Paleozoic potassium bentonites from Kinnekulle, Sweden: Clays and Clay Minerals, v. 34, p. 442–454.

Butts, Ch., 1940, Geology of the Appalachian Valley in Virginia: Virginia Geological Survey Bulletin 52, Part I, 568 p.

Butts, Ch., and Edmundson, R. S., 1943, Geology of the southwestern end of Walker Mountain, Virginia: Geological Society of America Bulletin, v. 54, p. 1669–1692.

Butts, Ch., and Gildersleeve, B., 1948, Geology and mineral resources of the Paleozoic area in northwest Georgia: Georgia Geological Survey Bulletin 54, 176 p.

Bystrom, A. M., 1956, Mineralogy of the Ordovician bentonite beds at Kinnekulle Sweden: Sveriges Geologiska Undersokning, series C, no. 540, 62 p.

Carter, B. D., and Chowns, T. M., 1986, Stratigraphic and environmental relationships of Middle and Upper Ordovician rocks in northwestern Georgia and northeastern Alabama, *in* Benson, D. J., and Stock, C. W., eds., Depositional history of the Middle Ordovician of the Alabama Appalachians (23rd Annual Field Trip Guidebook): Tuscaloosa, Alabama Geological Society, p. 33–50.

Chowns, T. M., 1986, A reinterpretation of the Rome, Helena, and Gadsden faults in Alabama: A preliminary report, *in* Benson, D. J., and Stock, C. W., eds., Depositional history of the Middle Ordovician of the Alabama Appalachians (23rd Annual Field Trip Guidebook): Tuscaloosa, Alabama Geological Society, p. 51–60.

Chowns, T. M., and Carter, B. D., 1983a, Stratigraphy of Middle and Upper Ordovician red beds in Georgia, *in* Chowns, T. M., ed., Geology of Paleozoic rocks in the vicinity of Rome, Georgia (18th Annual Field Trip Guidebook): Atlanta, Georgia Geological Society, p. 1–15.

Chowns, T. M., and Carter, B. D., 1983b, Middle Ordovician section along Mount Alto Road at southwest end of Horseleg Mountain, *in* Chowns, T. M., ed., Geology of Paleozoic rocks in the vicinity of Rome, Georgia (18th Annual Field Trip Guidebook): Atlanta, Georgia Geological Society, p. 70–73.

Chowns, T. M., and McKinney, F. K., 1980, Depositional facies in Middle-Upper Ordovician and Silurian rocks of Alabama and Georgia, *in* Frey, R. W., and Neathery, T. N., eds., Excursions in southeastern geology, v. II, Field trip guidebook for the annual meeting of the Geological Society of America, Field Trip no. 16, p. 323–348.

Coker, A. E., 1962, A mineralogical study of an Ordovician metabentonite near Clinton, Anderson County, Tennessee [M.S. thesis]: Knoxville, University of Tennessee, 49 p.

Conkin, J. E., Cerrito, P., and Kubacko, J., Jr., 1992a, Methods of identification of weathered wind-fall pyroclastics and their significance in correlation: Geological Society of America Abstracts with Programs, Northeast Section, v. 24, p. 13.

Conkin, J. E., Kubacko, J., Jr., Cerrito, P., and Conkin, B. M., 1992b, Middle Ordovician (Mohawkian) paracontinuous stratigraphy and air-fall pyroclastics in New York and southern Ontario: Geological Society of America Abstracts with Programs, Northeast Section, v. 24, p. 13.

Cressler, C. W., 1970, Geology and ground-water resources of Floyd and Polk Counties, Georgia: Georgia Geological Survey Information Circular 39, 95 p.

Cressman, E. R., 1973, Lithostratigraphy and depositional environments of the Lexington Limestone (Ordovician) of central Kentucky: U.S. Geological Survey Professional Paper 768, 61 p.

Cressman, E. R., and Noger, M. C., 1976, Tidal-flat carbonate environments in the High Bridge Group (Middle Ordovician) of central Kentucky: Kentucky Geological Survey Series X, Report of Investigations 18, 15 p.

Cullen-Lollis, J., and Huff, W. D., 1986, Correlation of Champlainian (Middle Ordovician) K-bentonite beds in central Pennsylvania based on chemical fingerprinting: Journal of Geology, v. 94, p. 865–874.

Delano, J. W., and 7 others, 1990, Petrology and geochemistry of Ordovician K-bentonites in New York State: Constraints on the nature of a volcanic arc: Journal of Geology, v. 98, p. 157–170.

Dennison, J. M., 1991, Sea level drop contrasted with peripheral bulge model for Appalachian basin during Mid-Ordovician: Geological Society of America Abstracts with Programs Northeast-Southeast Sections, v. 23, p. 21.

Dennison, J. M., Bambach, R. K., Dorobek, S. L., Filer, J. K., and Shell, J. A., 1992, Silurian and Devonian unconformities in southwestern Virginia, *in* Dennison, J. M., and Stewart, K. G., eds., Geologic field guides to North Carolina and vicinity: University of North Carolina at Chapel Hill, Department of Geology Geologic Guidebook no. 1, Field Trip no. 7, p. 79–114.

Desborough, G. A., Pitman, J. K., and Donnell, J. R., 1973, Microprobe analysis of biotites—A method of correlating tuff beds in the Green River Formation, Colorado and Utah: U.S. Geological Survey Journal of Research, v. 1, p. 39–44.

Diecchio, R. J., 1985, Post-Martinsburg Ordovician stratigraphy of Virginia and West Virginia: Virginia Division of Mineral Resources Publication 57, 77 p.

Dokken, K., 1987, Trace fossils from Middle Ordovician Platteville Formation, *in* Sloan, R. E., ed., Middle and Late Ordovician lithostratigraphy and biostratigraphy of the Upper Mississippi Valley: Minnesota Geological Survey Report of Investigations 35, p. 191–196.

Drahovzal, J. A., and Neathery, T. L., eds., 1971, The Middle and Upper Ordovician of the Alabama Appalachians (9th Annual Field Trip Guidebook): Tuscaloosa, Alabama Geological Society, 240 p.

Duane, M. J., and de Wit, M. J., 1988, Pb-Zn ore deposits of the northern Caledonides: Products of continental-scale fluid mixing and tectonic expulsion during continental collision: Geology, v. 16, p. 999–1002.

Dypvik, H., 1983, Clay mineral transformations in Tertiary and Mesozoic sediments from North Sea: American Association of Petroleum Geologists Bulletin, v. 67, p. 160–165.

Elliott, W. C., and Aronson, J. L., 1987, Alleghanian episode of K-bentonite illitization in the southern Appalachian Basin: Geology, v. 15, p. 735–739.

Elliott, W. C., and Aronson, J. L., 1993, The timing and extent of illite formation in Ordovician K-bentonites at the Cincinnati Arch, the Nashville Dome, and north-eastern Illinois basin: Basin Research, v. 5, p. 125–135.

Elliott, W. C., Aronson, J. L., Matisoff, G., and Gautier, D. L., 1991, Kinetics of the smectite to illite transformation in the Denver Basin: Clay mineral, K-Ar data, and mathematical model results: American Association of Petroleum Geologists Bulletin, v. 75, p. 436–462.

Epstein, A. G., Epstein, J. B., and Harris, L. D., 1977, Conodont color alteration—An index to organic metamorphism: U.S. Geological Survey Professional Paper 995, 27 p.

Fisher, R. S., and Land, L. S., 1986, Diagenetic history of Eocene Wilcox sandstones, south-central Texas: Geochimica Cosmochimica Acta, v. 50, p. 551–561.

Fisher, R. V., and Schmincke, H.-U., 1984, Pyroclastic rocks: Berlin, Springer-Verlag, 472 p.

Forsman, N. F., 1984, Discussion *of* "Misuse of the term 'bentonite' for ash beds of Devonian age in the Appalachian basin": Geological Society of America Bulletin, v. 95, p. 124.

Fox, P. P., and Grant, L. F., 1944, Ordovician bentonites in Tennessee and adjacent states: Journal of Geology, v. 52, p. 319–332.

Gold, P. B., 1987, Textures and geochemistry of authigenic albite from Miocene sandstones, Louisiana Gulf Coast: Journal of Sedimentary Petrology, v. 57, p. 353–362.

Grover, G., Jr., and Read, J. F., 1983, Paleoaquifer and deep burial related cements defined by regional cathodoluminescent patterns, Middle Ordovician carbonates, Virginia: American Association of Petroleum Geologists Bulletin, v. 67, p. 1275–1303.

Günal, A., 1979, Clay mineralogy, petrography, chemical composition, and stratigraphic correlation of some Middle Ordovician K-bentonites in the eastern mid-continent [Ph.D. dissertation]: Cincinnati, Ohio, University of Cincinnati, 237 p.

Gutstadt, A. M., 1958, Cambrian and Ordovician stratigraphy and oil and gas possibilities in Indiana: Indiana Department of Conservation, Geological Survey Bulletin 14, 103 p.

Hall, J. C., 1986, Conodonts and conodont biostratigraphy of the Middle Ordovician in the western overthrust region and Sequatchie Valley of the southern Appalachians [Ph.D. dissertation]: Columbus, Ohio State University, 345 p.

Hall, J. C., Bergstrom, S. M., and Schmidt, M. A., 1986, Conodont biostratigraphy of the Middle Ordovician Chickamauga Group and related strata of the Alabama Appalachians, *in* Benson, D. J., and Stock, C. W., eds., Depositional history of the Middle Ordovician of the Alabama Appalachians (23rd Annual Field Trip Guidebook): Tuscaloosa, Alabama Geological Society, p. 61–80.

Hamilton, W. B., 1979, Tectonics of the Indonesian region: U.S. Geological Survey Professional Paper 1078, 345 p.

Handwerk, R. H., 1981, Basin analysis of upper Middle Ordovician strata in southwestern Virginia and northeastern Tennessee [M.S. thesis]: Athens, Ohio University, 118 p.

Harris, A. G., Harris, L. D., and Epstein, J. B., 1978, Oil and gas data from Paleozoic rocks of the Appalachian basin: maps for assessing hydrocarbon potential and thermal maturity (conodont color alteration isograds and overburden isopachs): U.S. Geological Survey Miscellaneous Geologic Investigations, Map I-917E.

Hasson, K. O., 1985, Devonian-Mississippian strata exposed in cuts along U.S. 25E, *in* Walker, K. R., ed., The geologic history of the Thorn Hill Paleozoic section (Cambrian-Mississippian), eastern Tennessee: University of Tennessee, Department of Geological Sciences Studies in Geology 10, p. 111–119.

Hatcher, R. D. Jr., and 8 others, 1992, Status report on the geology of the Oak Ridge Reservation: Oak Ridge National Laboratory Environmental Services Division Publication 3860, ORNL/TM-12074, 247 p.

Hay, R. L., 1962, Origin and diagenetic alteration of the lower part of the John Day Formation near Mitchell, Oregon, *in* Engel, A.E.J., James, H. L., and Leonard, B. F., eds., Petrologic studies: A volume to honor A. F. Buddington: Geological Society of America Buddington volume, p. 191–216.

Haynes, J. T., 1989, The mineralogy and stratigraphic setting of the Rocklandian (Upper Ordovician) Deicke and Millbrig K-bentonite Beds along the Cincinnati Arch and in the southern Valley and Ridge [Ph.D. dissertation]: Cincinnati, Ohio, University of Cincinnati, 234 p.

Haynes, J. T., 1991, Stratigraphy of the Waynesboro Formation (Lower and Middle Cambrian) near Buchanan, Botetourt County, Virginia: Virginia Division of Mineral Resources Publication 116, 22 p.

Haynes, J. T., 1992, Reinterpretation of Rocklandian (Upper Ordovician) K-bentonite stratigraphy in southwest Virginia, southeast West Virginia, and northeast Tennessee, with a discussion of the conglomeratic sandstones in the Bays and Moccasin Formations: Virginia Division of Mineral Resources Publication 126, 58 p.

Haynes, J. T., and Goggin, K. E., 1993, Field guide to the Ordovician Walker Mountain Sandstone Member: Proposed type section and other exposures: Virginia Minerals, v. 39, p. 25–37.

Haynes, J. T., and Huff, W. D., 1990, Discussion *of* "Origin and tectonic setting of Ordovician bentonites in North America: Isotopic and age constraints": Geological Society of America Bulletin, v. 102, p. 1439–1440.

Haynes, J. T., Huff, W. D., and Hay, R. L., 1987, Compositional variations in the Middle Ordovician Deicke (T-3) and Millbrig (T-4) K-bentonites in the southeastern United States: Geological Society of America Abstracts with Programs, North-Central Section, v. 19, p. 203.

Hearn, P. P., Jr., and Sutter, J. F., 1985, Authigenic potassium feldspar in Cambrian carbonates: Evidence of Alleghanian brine migration: Science, v. 228, p. 1529–1531.

Heiken, G., 1972, Morphology and petrology of volcanic ashes: Geological Society of America Bulletin, v. 83, p. 1961–1988.

Hergenroder, J. D., 1966, The Bays Formation (Middle Ordovician) and related rocks of the southern Appalachians [Ph.D. dissertation]: Blacksburg, Virginia Polytechnic Institute and State University, 325 p.

Hergenroder, J. D., 1973, Stratigraphy of the Middle Ordovician bentonites in the southern Appalachians: Geological Society of America Abstracts with Programs, v. 5, p. 403.

Hoffman, J., and Hower, J., 1979, Clay mineral assemblages as low grade metamorphic geothermometers: Application to the thrust faulted Disturbed Belt of Montana, U.S.A., *in* Scholle, P. A., and Schluger, P. R., eds, Aspects of diagenesis: SEPM Special Publication 26, p. 55–80.

Huff, W. D., 1963, Mineralogy of Ordovician K-bentonites in Kentucky, *in* Clays and clay minerals, Proceedings of the 11th National Conference: New York, Pergamon Press, p. 200–209.

Huff, W. D., 1983a, Discussion *of* "Misuse of the term 'bentonite' for ash beds of Devonian age in the Appalachian basin": Geological Society of America Bulletin, v. 94, p. 681–682.

Huff, W. D., 1983b, Correlation of Middle Ordovician K-bentonites based on chemical fingerprinting: Journal of Geology, v. 91, p. 657–669.

Huff, W. D., and Kolata, D. R., 1990, Correlation of the Ordovician Deicke and Millbrig K-bentonites between the Mississippi Valley and the southern Appalachians: American Association of Petroleum Geologists Bulletin, v. 74, p. 1736–1747.

Huff, W. D., and Türkmenoğlu, A. G., 1981, Chemical characteristics and origin of Ordovician K-bentonites along the Cincinnati Arch: Clays and Clay Minerals, v. 29, p. 113–123.

Huff, W. D., Whiteman, J. A., and Curtis, C. D., 1988, Investigation of a K-bentonite by x-ray powder diffraction and analytical transmission electron microscopy: Clays and Clay Minerals, v. 36, p. 83–93.

Huff, W. D., Bergström, S. M., and Kolata, D. R., 1992, Gigantic Ordovician volcanic ash fall in North America and Europe: Biological, tectonomagmatic, and event-stratigraphic significance: Geology, v. 20, p. 875–878.

Huffman, G. G., 1945, Middle Ordovician limestones from Lee County, Virginia, to central Kentucky: Journal of Geology, v. 53, p. 145–174.

Janusson, V., and Bergström, S. M., 1980, Middle Ordovician faunal spatial differentiation in Baltoscandia and the Appalachians: Alcheringa, v. 4, p. 89–110.

Jenkins, C. M., 1984, Depositional environments of the Middle Ordovician Greensport Formation and Colvin Mountain Sandstone in Calhoun, Etowah, and St. Clair Counties, Alabama [M.S. thesis]: Tuscaloosa, University of Alabama, 156 p.

Johnsson, M. J., 1986, Distribution of maximum burial temperatures across northern Appalachian Basin and implications for Carboniferous sedimentation patterns: Geology, v. 14, p. 384–387.

Kastner, M., and Siever, R., 1979, Low temperature feldspars in sedimentary rocks: American Journal of Science, v. 279, p. 435–479.

Kay, G. M., 1931, Stratigraphy of the Ordovician Hounsfield metabentonite: Journal of Geology, v. 39, p. 361–370.

Kay, G. M., 1935, Distribution of Ordovician altered volcanic materials and related clays: Geological Society of America Bulletin, v. 46, p. 225–244.

Kay, G. M., 1944, Middle Ordovician of central Pennsylvania: Journal of Geology, v. 52, p. 1–23 and 97–116.

Kay, G. M., 1956, Ordovician limestones in the western anticlines of the Appalachians in West Virginia and Virginia northeast of the New River: Geological Society of America Bulletin, v. 67, p. 55–106.

Kerr, P. F., and Kulp, J. L., 1949, Reference clay localities—United States: American Petroleum Institute Project 49, Clay Mineral Standards, Preliminary Report no. 2, 103 p.

Kesler, S. E., Jones, L. M., and Ruiz, J., 1988, Strontium isotopic geochemistry of Mississippi Valley–type deposits, East Tennessee: Implications for age and source of mineralizing brines: Geological Society of America Bulletin, v. 100, p. 1300–1307.

Kolata, D. R., Huff, W. D., and Frost, J. K., 1984, Correlation of Champlainian (Middle Ordovician) K-bentonites from Minnesota to Kentucky and Tennessee: Geological Society of America Abstracts with Programs, v. 16, p. 563.

Kolata, D. R., Frost, J. K., and Huff, W. D., 1986, K-Bentonites of the Ordovician Decorah Subgroup, Upper Mississippi Valley: Correlation by chemical fingerprinting: Illinois State Geological Survey Circular 537, 30 p.

Kreisa, R. D., 1980, The Martinsburg Formation (Middle and Upper Ordovician) and related facies in southwestern Virginia [Ph.D. dissertation]: Blacksburg, Virginia, Virginia Polytechnic Institute and State University, 335 p.

Kunk, M. J., and Sutter, J. F., 1984, $^{40}Ar/^{39}Ar$ age spectrum dating of biotites from Middle Ordovician bentonites, eastern North America, *in* Bruton, D. L., ed., Aspects of the Ordovician System: Palaeontological Contributions from the University of Oslo, no. 295, p. 11–22.

Kunk, M. J., Sutter, J. F., and Bergstrom, S. M., 1984, $^{40}Ar/^{39}Ar$ age spectrum dating of biotite and sanidine from Middle Ordovician bentonites of Sweden: Comparison with results from eastern North America: Geological Society of America Abstracts with Programs, v. 16, p. 566.

Land, L. S., and Milliken, K. L., 1981, Feldspar diagenesis in the Frio Formation, Brazoria County, Texas Gulf Coast: Geology, v. 9, p. 314–318.

Lee, A. M., 1983, Lithofacies and depositional environments of the Chickamauga Group, Jefferson County, north-central Alabama [M.S. thesis]: Tuscaloosa, University of Alabama, 198 p.

Lesure, F. G., 1957, Geology of the Clifton Forge iron district, Virginia: Bulletin of Virginia Polytechnic Institute Engineering, Experiment Station Series 118, 130 p.

Lounsbury, R. W., and Melhorn, W. N., 1964, Clay mineralogy of Paleozoic K-bentonites of the eastern United States (part 1), *in* Clays and clay minerals, Proceedings of the 12th National Conference: New York, Pergamon Press, p. 557–565.

Lounsbury, R. W., and Melhorn, W. N., 1971, Clay mineralogy of some Ordovician Missouri and Virginia K-bentonites: Virginia Journal of Science, v. 22, p. 122.

Loutit, T. S., Hardenbol, J., Vail, P. R., and Baum, G. R., 1988, Condensed sections: The key to age determination and correlation of continental margin sequences, *in* Wilgus, C. K., Hastings, B. S., Kendall, C. G. St. C., Posamentier, H. W., Ross, C. A., and Van Wagoner, J. C., eds., Sea-level changes: An integrated approach: SEPM Special Publication no. 42, p. 183–213.

Lyons, P. C., Outerbridge, W. F., Congdon, R. D., Evans, H. T., Jr., and Slucher, E. R., 1992a, "Fingerprinting" of kaolinized volcanic ash beds (tonsteins) as a tool in the correlation of coal beds in the Appalachian basin: Geological Society of America Abstracts with Programs, Northeast Section, v. 24, p. 60.

Lyons, P. C., and 7 others, 1992b, An Appalachian isochron: A kaolinized Carboniferous air-fall volcanic-ash deposit (tonstein): Geological Society of America Bulletin, v. 104, p. 1515–1527.

Mack, G. H., 1985, Provenance of the Middle Ordovician Blount clastic wedge, Georgia and Tennessee: Geology, v. 13, p. 299–302.

Manley, F. H., and Martin, B. F., 1972, Clay mineralogy of some Ordovician bentonites from the Chickamauga Supergroup, northwest Georgia: Geological Society of America Abstracts with Programs, v. 4, p. 89-90.

Marshall, B. D., Woodard, H. H., and DePaolo, D. J., 1986, K-Ca-Ar systematics of authigenic sanidine from Waukau, Wisconsin, and the diffusivity of argon: Geology, v. 14, p. 936–938.

Maxwell, D. T., and Hower, J., 1967, High-grade diagenesis and low-grade metamorphism of illite in the Precambrian Belt Series: American Mineralogist, v. 52, p. 843–857.

Maynard, J. B., 1982, Extension of Berner's "New Geochemical Classification of Sedimentary Environments" to ancient sediments: Journal of Sedimentary Petrology, v. 52, p. 1325–1331.

McFarlan, A. C., 1943, Geology of Kentucky: Lexington, University of Kentucky, 531 p.

Milici, R. C., 1969, Middle Ordovician stratigraphy in central Sequatchie Valley, Tennessee: Southeastern Geology, v. 11, p. 111–128.

Milici, R. C., and Smith, J. W., 1969, Stratigraphy of the Chickamauga Supergroup in its type area: Georgia Geological Survey Bulletin 80, p. 1–35.

Milici, R. C., and Wedow, H., 1977, Upper Ordovician and Silurian stratigraphy in Sequatchie Valley and parts of adjacent Valley and Ridge, Tennessee: U.S. Geological Survey Professional Paper 996, 38 p.

Miller, R. L., and Brosgé, W. P., 1954, Geology and oil resources of the Jonesville district, Lee County, Virginia: U.S. Geological Survey Bulletin 990, 240 p.

Miller, R. L., and Englund, K. J., 1975, Geology of southwest Virginia coal fields and adjacent areas: Seventh Annual Virginia Geology Field Conference Guidebook, 34 p.

Miller, R. L., and Fuller, J. O., 1954, Geology and oil resources of the Rose Hill district—the fenster area of the Cumberland overthrust block—Lee County, Virginia: Virginia Geological Survey Bulletin 71, 383 p.

Mitchum, R. M, Jr., Vail, P. R., and Thompson, S., III, 1977, Seismic stratigraphy and global changes of sea level, Part 2: The depositional sequence as a basic unit for stratigraphic analysis, *in* Payton, C. E., ed., Seismic stratigraphy—Applications to hydrocarbon exploration: American Association of Petroleum Geologists Memoir 26, p. 53–62.

Moore, D. M., and Reynolds, R. C., Jr., 1989, X-ray diffraction and the identification and analysis of clay minerals: New York, Oxford University Press, 332 p.

Morad, S., 1988a, Albitized microcline grains of post-depositional and probable detrital origins in Brφttum Formation sandstones (Upper Proterozoic), Sparagmite Region of southern Norway: Geological Magazine, v. 125, p. 229–239.

Morad, S., 1988b, Diagenesis of titaniferous minerals in Jurassic sandstones from the Norwegian Sea: Sedimentary Geology, v. 57, p. 17–40.

Morad, S., and AlDahan, A. A., 1986, Alteration of detrital Fe-Ti oxides in sedimentary rocks: Geological Society of America Bulletin, v. 97, p. 567–578.

Morad, S., Bergan, M., Knarud, R., and Nystuen, J. P., 1990, Albitization of detrital plagioclase in Triassic reservoir sandstones from the Snorre Field, Norwegian North Sea: Journal of Sedimentary Petrology, v. 60, p. 411–425.

Morton, J. P., 1985, Rb-Sr evidence for punctuated illite/smectite diagenesis in the Oligocene Frio Formation, Texas Gulf Coast: Geological Society of America Bulletin, v. 96, p. 114–122.

Mussman, W. J., and Read, J. F., 1986, Sedimentology and development of a passive-to-convergent-margin unconformity: Middle Ordovician Knox unconformity, Virginia Appalachians: Geological Society of America Bulletin, v. 97, p. 282–295.

Neathery, T. L., and Drahovzal, J. A., 1986, Middle and Upper Ordovician stratigraphy of the southernmost Appalachians, *in* Neathery, T. L., ed., Geological Society of America Centennial Field Guide, Southeastern Section, v. 6, p. 167–172.

Nelson, W. A., 1921, Notes on a volcanic ash bed in the Ordovician of middle Tennessee: Tennessee Division of Geology Bulletin 25, p. 46–48.

Nelson, W. A., 1922, Volcanic ash bed in the Ordovician of Tennessee, Kentucky, and Alabama: Geological Society of America Bulletin, v. 33, p. 605–616.

Nelson, W. A., 1926, Volcanic ash deposit in the Ordovician of Virginia: Geological Society of America Bulletin, v. 37, p. 149–150.

Neuman, R. B., 1955, Middle Ordovician rocks of the Tellico-Sevier belt, eastern Tennessee: U.S. Geological Survey Professional Paper 274-F, p. F141–F178.

North American Commission on Stratigraphic Nomenclature, 1983, North American Stratigraphic Code: American Association of Petroleum Geologists Bulletin, v. 67, p. 841–875.

Oliver, J., 1986, Fluids expelled tectonically from orogenic belts: their role in hydrocarbon migration and other geologic phenomena: Geology, v. 14, p. 99–102.

Perry, E. A., and Hower, J., 1970, Burial diagenesis in Gulf Coast pelitic sediments: Clays and Clay Minerals, v. 18, p. 165–177.

Perry, W. J., Jr., 1964, Geology of the Ray Sponaugle well: American Association of Petroleum Geologists Bulletin, v. 48, p. 659–669.

Perry, W. J., Jr., 1972, The Trenton Group of Nittany Anticlinorium, eastern West Virginia: West Virginia Economic and Geological Survey Circular Series No. 13, 30 p.

Potter, P. E., Maynard, J. B., and Pryor, W. A., 1980, Sedimentology of shale: New York, Springer-Verlag, 306 p.

Rader, E. K., 1982, Valley and Ridge stratigraphic correlations, Virginia: Virginia Division of Mineral Resources Publication 37, 1 sheet.

Rader, E. K., and Gathright, T. M., 1986, Stratigraphic and structural features of Fincastle Valley and Eagle Rock Gorge, Botetourt County, Virginia: Geological Society of America, Centennial Field Guide, Southeastern Section, v. 6, p. 105–108.

Rader, E. K., and Perry, W. J., Jr., 1976, Reinterpretation of the geology of Brocks Gap, Rockingham County, Virginia: Virginia Minerals, v. 22, p. 37–45.

Raymond, D. E., 1978, Middle Ordovician conodonts from the Birmingham Expressway cut, Jefferson County, Alabama, *in* Kidd, J. T., and Shannon, S. W., eds., Stratigraphy and structure of the Birmingham area, Jefferson County, Alabama (16th Annual Field Trip Guidebook): Tuscaloosa, Alabama Geological Society, p. 44–67.

Read, J. F., 1980, Carbonate ramp-to-basin transition and foreland basin evolution, Middle Ordovician, Virginia Appalachians: American Association of Petroleum Geologists Bulletin, v. 64, p. 1575–1612.

Reade, E. H., Jr., 1959, Distinguishing characteristics of bentonites in northwestern Georgia: Bulletin of the Georgia Academy of Sciences, v. 17, p. 72–73.

Reynolds, R. C., 1980, Interstratified clay minerals, *in* Brindley, G. W., and Brown, G., eds., Crystal structures of clay minerals and their x-ray identification: Mineralogical Society of London, p. 249–303.

Reynolds, R. C., and Hower, J., 1970, The nature of interlayering in mixed-layer illite-montmorillonites: Clays and Clay Minerals, v. 18, p. 25–36.

Rodgers, J., 1953, Geologic map of east Tennessee with explanatory text: Tennessee Division of Geology Bulletin 58, Part II, 168 p.

Rose, W. I., and Chesner, C. A., 1987, Dispersal of ash in the great Toba eruption, 75 ka: Geology, v. 15, p. 913–917.

Rosenkrans, R. R., 1933, Bentonite in northern Virginia: Journal of the Washington Academy of Sciences, v. 23, p. 413–419.

Rosenkrans, R. R., 1936, Stratigraphy of Ordovician bentonite beds in southwestern Virginia, *in* Contributions to Virginia Geology—I: Virginia Geological Survey Bulletin 46-I, p. 85–111.

Ross, R. J., Jr., and 27 others, 1982a, The Ordovician system in the United States, correlation chart and explanatory notes: International Union of Geological Sciences, Publication 12, 1 sheet.

Ross, R. J., Jr., and 14 others, 1982b, Fission track dating of British Ordovician and Silurian stratotypes: Geological Magazine, v. 119, p. 135–153.

Samson, S. D., Kyle, P. R., and Alexander, E. C., Jr., 1988, Correlation of North American Ordovician bentonites by using apatite chemistry: Geology, v. 16, p. 444–447.

Samson, S. D., Patchett, P. J., Roddick, J. C., and Parrish, R. R., 1989, Origin and tectonic setting of Ordovician bentonites in North America: Isotopic and age constraints: Geological Society of America Bulletin, v. 101, p. 1175–1181.

Scarfe, C. M., Fujii, T., and Harris, D. M., 1982, Mineralogy and geochemistry of pumice from 19 March 1982 eruption of Mount St. Helens: Geological Society of America Abstracts with Programs, v. 14, p. 608.

Scharpf, C. D., 1990, Stratigraphy and associated faunas of the Middle Ordovician (Rocklandian) Millbrig K-bentonite in central Kentucky [M.S. thesis]: Cincinnati, Ohio, University of Cincinnati, 280 p.

Schmidt, M. A., 1982, Conodont biostratigraphy and facies relations of the Chickamauga Limestone (Middle Ordovician) of the southern Appalachians, Alabama and Georgia [M.S. thesis]: Columbus, Ohio State University, 270 p.

Self, S., Rampino, M. R., Newton, M. S., and Wolff, J. A., 1984, Volcanological study of the great Tambora eruption of 1815: Geology, v. 12, p. 659–663.

Shaver, R. H., regional coordinator, 1985, Midwestern basin and arches region: Tulsa, American Association of Petroleum Geologists, Correlation of Stratigraphic Units of North America (COSUNA) Project, 1 map.

Simonson, J.C.B., 1985, Mixed carbonate-siliciclastic tidal flat deposits of the Moccasin Formation, *in* Walker, K. R., ed., The geologic history of the Thorn Hill Paleozoic section (Cambrian-Mississippian), eastern Tennessee: University of Tennessee, Department of Geological Sciences Studies in Geology 10, p. 76–85.

Sloan, R. E., 1987, Introduction to the Middle and Late Ordovician field trips, *in* Balaban, N. H., ed., Field trip guidebook for the Upper Mississippi Valley—Minnesota, Iowa, and Wisconsin: Minnesota Geological Survey Guidebook Series 15, p. 45–52.

Sloan, R. E., 1988, The Deicke ash bed and the Blackriver/Trenton boundary: Geological Society of America Abstracts with Programs, North-Central Section, v. 20, p. 89.

Snäll, S., 1977, Silurian and Ordovician bentonites of Gotland (Sweden): Acta University Stockholm, Contributions in Geology 31, 80 p.

Sparks, R.S.J., Self, S., and Walker, G.P.L., 1973, Products of ignimbrite eruptions: Geology, v. 1, p. 115–118.

Środoń, J., and Eberl, D. D., 1984, Illite, *in* Bailey, S. W., ed., Micas: Mineralogical Society of America Reviews in Mineralogy, v. 13, p. 495–544.

Sweet, W. C., 1984, Graphic correlation of upper Middle and Upper Ordovician rocks, North American midcontinent province, U.S.A., *in* Bruton, D. L., ed., Aspects of the Ordovician System: Palaeontological Contributions from the University of Oslo, no. 295, p. 23–35.

Sweet, W. C., and Bergstrom, S. M., 1974, Provincialism exhibited by Ordovician conodont faunas, *in* Ross, C. A., ed., Paleogeographic provinces and provinciality: Society of Economic Paleontologists and Mineralogists Special Publication 21, p. 189–202.

Thomas, W. A., 1982, Stratigraphy and structure of the Appalachian fold and thrust belt in Alabama, *in* Thomas, W. A., and Neathery, T. L., eds., Appalachian thrust belt in Alabama: Tectonics and sedimentation (Guidebook for Field Trip 13, Geological Society of America Annual Meeting, New Orleans): Tuscaloosa, Alabama Geological Society, Tuscaloosa, p. 55–66.

Thomas, W. A., Tull, J. F., Neathery, T. L., Mack, G. H., and Ferrill, B. A., 1982, A field guide to the Appalachian thrust belt in Alabama, *in* Thomas, W. A., and Neathery, T. L., eds., Appalachian thrust belt in Alabama: Tectonics and sedimentation (Guidebook for Field Trip 13, Geological Society of America Annual Meeting, New Orleans): Tuscaloosa, Alabama Geological Society, p. 5–40.

Ulrich, E. O., 1911, Revision of the Paleozoic systems: Geological Society of America Bulletin, v. 22, p. 281–680.

Van Wagoner, J. C., and 6 others, 1988, An overview of the fundamentals of sequence stratigraphy and key definitions, *in* Wilgus, C. K., Hastings, B. S., Kendall, C. G. St. C., Posamentier, H. W., Ross, C. A., and Van Wagoner, J. C., eds., Sea-level changes: An integrated approach: Society of Economic Paleontologists and Mineralogists Special Publication 42, p. 39–45.

Walker, K. R., and Diehl, W. W., 1985, Upper Ordovician "Martinsburg" Formation, a deep-water to shallow-water sequence, *in* Walker, K. R., ed., The geologic history of the Thorn Hill Paleozoic section (Cambrian-Mississippian), eastern Tennessee: University of Tennessee, Department of Geological Sciences Studies in Geology 10, p. 86–92.

Walker, T. R., 1984, 1984 SEPM Presidential Address: Diagenetic albitization of potassium feldspar in arkosic sandstones: Journal of Sedimentary Petrology, v. 54, p. 3–16.

Walker, T. R., Waugh, B., and Grone, A. J., 1978, Diagenesis in first-cycle desert alluvium of Cenozoic age, southwestern United States and northwestern Mexico: Geological Society of America Bulletin, v. 89, p. 19–32.

Ward, W. I., 1983, Lithofacies and depositional environments of a portion of the Stones River Formation in Etowah and DeKalb Counties, northeast Alabama [M.S. thesis]: Tuscaloosa, University of Alabama, 255 p.

Watanabe, T., 1981, Identification of illite/montmorillonite interstratifications by x-ray powder diffraction: Journal of the Mineralogical Society of Japan, Special Issue 15, p. 32–41.

Waugh, B., 1978, Authigenic K-feldspar in British Permo-Triassic Sandstones: Journal of the Geological Society of London, v. 135, p. 51–56.

Weaver, C. E., 1953, Mineralogy and petrology of some Ordovician K-bentonites and related limestones: Geological Society of America Bulletin, v. 64, p. 921–944.

Weaver, C. E., and Beck, K. C., 1971, Clay-water diagenesis during burial: How mud becomes gneiss: Geological Society of America Special Paper 134, 96 p.

Westgate, J. A., and Fulton, J. J., 1975, Tephra-stratigraphy of Olympia interglacial sediments in south-central British Columbia, Canada: Canadian Journal of Earth Sciences, v. 12, p. 489–502.

Willman, H. B., and Kolata, D. R., 1978, The Platteville and Galena Groups in northern Illinois: Illinois State Geological Survey Circular 502, 75 p.

Wilson, C. W., Jr., 1949, Pre-Chattanooga stratigraphy in central Tennessee: Tennessee Division of Geology Bulletin 56, 407 p.

Wilson, C. W., Jr., 1991, The Geology of Nashville, Tennessee: Tennessee Division of Geology, Bulletin 53 2nd ed., 185 p.

Woodrow, D. L., Dennison, J. M., Ettensohn, F. R., Sevon, W. T., and Kirchgasser, W. T., 1988, Middle and Upper Devonian stratigraphy and paleogeography of the central and southern Appalachians and eastern Midcontinent, *in* McMillan, N. J., Embry, A. F., and Glass, D. J., eds., Devonian of the world: Canadian Society of Petroleum Geologists Memoir 14, v. 1, p. 277–301.

Woodward, H. P., 1951, Ordovician system of West Virginia: West Virginia Economic and Geological Survey, v. 21, 627 p.

Yen, F., and Goodwin, J. H., 1976, Correlation of tuff layers in the Green River Formation, Utah, using biotite compositions: Journal of Sedimentary Petrology, v. 46, p. 345–354.

Young, D.M., 1940, Bentonitic clay horizons and associated chert layers of central Kentucky: University of Kentucky Research Club Bulletin, v. 6, p. 27–31.

Zhang, Y., and Huff, W. D., 1992, Grain size as an indicator of ash source and distribution in Middle Ordovician K-bentonites of eastern North America: Geological Society of America Abstracts with Programs, v. 24, p. A267.

Zhenzhong, G., and Eriksson, K. A., 1991, Internal-tide deposits in an Ordovician submarine channel: Previously unrecognized facies? Geology, v. 19, p. 734–737.

Manuscript Accepted by the Society August 18, 1993

Typeset by WESType Publishing Services, Inc., Boulder, Colorado
Printed in U.S.A. by Johnson Printing, Boulder, Colorado